RELIGION AND ECOLOGY

RELIGION AND ECOLOGY

developing a planetary ethic

whitney a. bauman

COLUMBIA UNIVERSITY PRESS   New York

COLUMBIA UNIVERSITY PRESS
PUBLISHERS SINCE 1893
NEW YORK   CHICHESTER, WEST SUSSEX

cup.columbia.edu

Copyright © 2014 Columbia University Press

All rights reserved

Library of Congress Cataloging-in-Publication Data
Bauman, Whitney.
    Religion and ecology: developing a planetary ethic / Whitney A. Bauman.
        pages cm
    Includes bibliographical references and index.
    ISBN 978-0-231-16342-2 (cloth)—ISBN 978-0-231-16343-9 (pbk.)—
ISBN 978-0-231-53710-0 (e-book)
    1. Ecology—Religious aspects.   I. Title.
    BL65.E36B38 2014
    201′.77—dc23
                                                            2013028197

COVER IMAGE: 84 MILLION STARS IN MILKY WAY, © OHAINAUT/ESO/
HANDOUT/DPA/CORBIS
COVER DESIGN: CHANG JAE LEE

CONTENTS

Acknowledgments   vii

Introduction: The Emergence of Planetary Identities   1

1. Religion and Science in Dialogue   17

    A New Version of Agrippa's Trilemma: Toward Planetary Knowing   18

    Defining the Terms:
    Religion, Science, and Nature in a Planetary Context   22

    The Ptolemaic Cosmology, the Scientific "Revolution," and the Grounds for Globalization   26

    The Myth of the Secular and the Space of Capitalism:
    Globalatinization   31

2. Destabilizing Nature:
    Natura Naturans, Emergence, and Evolution's Rainbow   37

    Bruno, Spinoza, and the Emergentists:
    Stories of Radical Immanence   42

    Radical Materialism, Pragmatism, and Habits of Nature-Culture   48

    Nonequilibrium Thermodynamics and Open Evolving Systems   54

    Queering Nature   56

3. Destabilizing Religion: The Death of God, a Viable Agnosticism, and the Embrace of Polydoxy   63

    Unknowing at the Edges: Making Meaning in the Dark   64

    Emergent Meaning-Making Practices: Toward a Viable Agnosticism   71

    A Call for Planetary Religions: Our Contexts for Meaning Making   75

    Religions as Lines of Flight: Embracing Polydoxy   82

4. Destabilizing Identity: Beyond Identity Solipsism   85

   The Identity Trilemma: Identity Solipsism and the Colonial "I"   86

   The Capitalist Politics of Identity Construction   102

5. The Emergence of Ecoreligious Identities   107

   Technologies of Meaning   109

   Transcultural Ecoreligious Identities   119

   Performing Meaning:
   Taking on the Abject Toward Planetary Identities   123

6. Developing Planetary Environmental Ethics:
   A Nomadic Polyamory of Place   127

   Planetary Flows: A Context of Movement   128

   Slippery Slopes: The Epistem-ontology of Planetary Ethics   130

   The Ethical/Religious Level: Common Grounds   135

   The Anthropological/Political Level: Uncharted Territories   140

   Components of a Planetary Environmental Ethic   144

7. Challenging Human Exceptionalism:
   Human Becoming, Technology, Earth Others, and
   Planetary Identities   151

   Becoming Plant, Mineral, Animal   154

   We Are Hybrid, Cyborg, Biohistorical   159

   Hyperobjectivity/Hypersubjectivity   161

   Challenging Agency, Hypocrisy, Integrity, and Love   164

Notes   173
Glossary   193
Works Cited   217
Index   227

ACKNOWLEDGMENTS

As with any written project, this one represents the manifestation of many biohistorical flows. My ideas, like all ideas, are never fully my own and are fed by the histories of thinking in many different cultures throughout the world. Likewise, they are fed by the elements of the 13.7-billion-year process of cosmic expansion and 4.5 billions of years of geoevolution. Thinking more locally, there are many other earth bodies at whose expense I think, because food, water, air, and technologies are necessary in order for me to continue living on as a centralized response-able organism. So, I have to start these pages by thanking these planetary others: human, animal, plant, mineral, machine, and the processes that have led to the manifestation of my own subjective becoming.

More formally, this book would not be possible without the help of several academic influences over the past five years. I first want to thank all of my friends and colleagues at Florida International University, where I have been fortunate and privileged enough to be on a tenure-track position in religious studies. This privileged economic space is essential for the completion of such a project. Second, I want to thank all the organizations that have nurtured my thinking in these areas: the Religion and Ecology Group of the American Academy of Religion; the Forum on Religion and Ecology; the International Society for the Study of Religion, Nature, and Culture; the Queer Theory Reading Group at the University of Miami; InterFaithOut, a Project of

Save Dade; and the Institute for Religion in an Age of Science are the primary arenas that have been fruitful for my thinking.

This project was also made possible by a number of grants and fellowships, including the 2009 Templeton Award for Theological Promise, FIU College of Arts and Sciences Summer Research Grant 2009, Wabash Summer Research Fellowship 2010, the Bhagwan Mahavir Junior Faculty Fellowship 2011, the Morris and Anita Broad Research Fellowship 2011, and a USINDO Travel Grant 2012. It was in Heidelberg in 2009 at conference associated with the Templeton Award for Theological Promise that these ideas got off to a very rocky start, and I thank all of the participants there, especially Michael Welker and Peter Paris, for the brutal honesty that pushed me into thinking about these ideas in a much more sustained way. The Bhagwan Mahavir and Broad Fellowships allowed me to travel through India in the summer of 2011, visiting various Jain communities and researching "Jainism and science." Some of the ideas found here about multiperspectivalism are in deep debt to that experience. Furthermore, I have been fortunate enough to travel to Indonesia and teach at the University of Gadjah Mada in Yogyakarta, Indonesia several summers now and the dialogues I have begun there with friends and colleagues have contributed greatly to these pages.

In addition to these academic habitats and sources of funding, a dozen other conference papers and countless conversations at those conferences have led to the developments herein. They are too long to enumerate here, but I will mention several publications that were instrumental for developing these ideas, which emerged from some of these conferences. Chapter 1 of this text is partially developed in an article that I wrote for the Center for Civilizational Dialogue at the University of Malaysia and that appears as "Religion, Science, and Nature: Shifts in Meaning on a Changing Planet" in *Zygon: Journal of Religion and Science* 46, no. 4 (December 2011): 777–792. The ideas in chapter 3 were first developed in a conference on "Energy and God" at the University of Central Arkansas in Conway and appeared as "Emergence, Energy, and Openness: A Viable Agnostic Theology" in *Cosmology, Ecology and the Energy of God*, ed. Donna Bowman and Clayton Crockett (Fordham University Press, 2011), 70–84. Finally, the ideas of the identity trilemma found in chapter 4 were first developed through my chapter in *Voices of Feminist Liberation: Writings in Celebration of Rosemary Radford Ruether*, ed. Emily Leah Silverman, Dirk von der Horst, and Whitney Bauman (Equinox, 2012). Though each of them is much modified in the current text, these three pieces greatly contributed to my overall thinking about "planetary identities." Several other articles and chapters have of course shaped my writing along the way over the past few years, including:

"De-Constructing Transcendence: The Emergence of Religious Bodies" (with James W. Haag) in *The Body and Religion: Modern Science and the Construction of Religious Meaning*, ed. David Cave and Rebecca Norris (Brill 2012), 37–55.

Inherited Land: The Changing Grounds of Religion and Ecology, ed. Whitney Bauman, Richard Bohannon, and Kevin O'Brien (Wipf and Stock, 2011).

"Fashioning a Persuasive Environmental Ethic: Thinking Without Surface and Depth" in *Ecozona* 2, no. 2 (2011): 17–39.

"Technology and the Polytheistic Mind: From the Truth of the Global to Planetary 'Lines of Flight'" in *Dialog: A Journal of Theology* 50, no. 4 (Winter 2011): 344–353.

"Christianity and Nature" in the *Routledge Companion for Religion and Science*, ed. Gregory Peterson, Michael Spezio, and James W. Haag (Routledge, 2011), 368–378.

These writings in particular are important for the ideas weaving in and out of the pages of this book. I thank all of the editors, coeditors, and coauthors I worked with on these projects, especially Rick Bohannon, Kevin O'Brien, James Haag, Emily Silverman, and Dirk von der Horst.

In addition to these formal interlocutors in writing and conferencing, I must also think a number of friends and colleagues who have helped me bounce these ideas around: Christine Gudorf, Laurie Schrage, Jose Gabilando, Steven Blevins, Brenna Munroe, and Catherine Keller. Through numerous conversations and what may seem like informal encounters, these people have especially helped shape the ideas of this text. I am also indebted to the anonymous reviewers of this manuscript and of course to my editors at Columbia University Press, Wendy Lochner and Christine Dunbar. A huge thanks goes to Kimberly Zwez, my graduate student assistant, for painfully reading through versions of this text and compiling the list of terms for the glossary. For everything that you can understand in these pages, it is as the result in no small part of Kimberly's clarifications; for those things that are still obtuse, blame them on me. I also give special thanks to Christine Gudorf, who proofread and commented on the final version of this manuscript for me: thanks for making this a better book and sorry it will inevitably fall short of your perfection! In addition, I would be remiss if I did not thank my closest friends and family for supporting me, loving me, making me laugh, and making me get out of my head: Immy, Mom, Dad, Jim, Ceciley, Andrew,

Marshall, Little D, Steven, Ms. Terry, Bob, Giorgio, Matt, and Nick—I truly love you all. Finally, I have to thank the students I have had over the past five years at Florida International University and at the University of Gadjah Mada. Teaching has refined my thinking, and the classroom has been a source of fireworks for my mind; without all these students this book would definitely not be possible. It is to them that I dedicate this book.

RELIGION AND ECOLOGY

INTRODUCTION
The Emergence of Planetary Identities

> An occurrence or event, in fact, is perhaps nothing other than an interstitial crossing of multiplicities, whose difference, decision, the spacing of differentiation, chance, and the uncertain nothing, all meet in a moment of becoming.
> MICHAEL ANKER, *The Ethics of Uncertainty*

This book is an attempt to think about new ways of understanding self-and-other beyond substance-based notions of identity. Substance-based notions of identity work on the assumption that all things have essential components to which they can be reduced, whether these components be material (genes) or immaterial (a soul or mind). Substance-based identities rely also on a metaphysic of substance that claims reality can be reduced to some smallest common denominator. As such, the book begins from questions of how to think self, other, and difference without the narrated markers of substance-based identity boundaries. In other words, it addresses the question of how we maintain the difference necessary for multiplicity without reifying difference into conceptual isolations that ignore the continuous becoming (other) of all life on the planet. Such emergent, phenomenal, and event-based understandings of identity and reality challenge histories that narrate identity origins and any knowledge claims based on "origins," or foundational or reductive ways of thinking. This focus on the (co)construction of identities and knowledge is, of course, not new to religious studies, philosophy, or the humanities.

For many years now, liberal arts colleges and universities in Europe and the United States (at least) have been experimenting with new ways to teach history: the history of ideas, cultures, thought, religions, science, etc. The old "Western civilization" course just doesn't do the job anymore, as we become more and more aware of the multiple histories, stories, sciences, and interactions between geographies, cultures, religions, and civilizations throughout these histories. In a sense, we are beginning to realize our "hybridity."[1] This hybridity does not only entail the hybridity of histories and cultures, those things in the mythical land called "the West" that we place on the human side of the human/nature divide. Rather, they also include hybridity with other species and technology.[2] Thus this book is nothing more than an attempt to think about this hybridity and what it means for our own identities and our ethical obligations toward other humans and planetary beings. As such, writing an introduction that begins from multiple starting places, multiple identities, multiple realms of knowledge is a struggle, to say the least. Whatever starting point I begin with will privilege a certain identity, lens, or perspective on the explorations that take place between the beginnings and endings of this book. So I begin here with several different beginnings as a way to bring the reader into the exploration of ideas that fall within the horizons of this book (always already a snapshot of thought that must make epistemic and narrative cuts in order to begin, end, and then hopefully begin again). These beginnings are really meant to stimulate thought, imaginative adventures, and to further conversations rather than secure any definite knowledge or conclusions. Such exercises are deeply important in a world that is increasingly reified into conceptually bounded ways of thinking and corresponding identities and interactions. This "conceptual boundedness" is reflected, for instance, in the economization of higher education, where the space for critical and creative thinking and learning is more and more eroded by the need for a university or college to justify an economic bottom line or for students to think of their degrees as a step along the way to a career. This economization of thought also marks the process of globalization. In fact, as I will argue, the conceptual violence created by imposing sameness over the face of the globe through a process of the globalization of Western ways of thinking is indeed the result of conceptually bounded identities and knowledge. By conceptual violence, I mean that forcing life into specific categories of production, politics, and identity (what is "male" and "female" for instance), leaves out many alternatives and in fact ends up creating violence toward alternative embodiments. Think of the erosion of "common lands" that takes place when countries open up to economic globalization and the corresponding increase in poverty (starving bodies), all due to the fact that production must conform to capitalistic modes of production. Or, think of the abuse toward bodies that happens to trans identities when Western ways of thinking of gender and sex are imposed upon other cultures that have traditionally held spaces for "third genders" and "third sexes." Finally, consider what happens when Western understandings that separate humans from the rest of the natural world are imposed upon communities where landscape and culture are inextricably intertwined (leading to cultural death for the human communities and resource extraction and threatened species for the more-than-human community). In thinking about these, among other issues, this book is hopefully a catalyst for thinking anew with the rest of the planetary community toward a future becoming that is radically unknown. It is not that we ought not use concepts. Indeed as meaning-making creatures we are bound to conceptual thinking. It is, rather, that we ought not confuse our concepts with reality or even with the best way of thinking about the worlds in which we live, which themselves cocreate us.

As Foucault notes, "Each society has its regime of truth, its 'general politics' of truth: that is, the types of discourse which it accepts and makes function as true; the mechanisms and instances which enable one to distinguish true and false statements, the means by which each is sanctioned; the techniques and procedures accorded value in the acquisition of truth; the status of those who are charged with saying what counts as true."[3] Producing such knowledge regimes takes material and energy resources. In this book I am suggesting that the regime of truth that might be identified as "globalization" (which is the heir to Enlightenment and modern regimes) has now run up against planetary boundaries. In other words, the weight and costs of the technologies that perpetuate this type of universal and universalizing truth is outstripping the planets' very capacity to reproduce itself.[4] As such, this book is an attempt to begin thinking about new knowledge regimes that foster planetary rather than just (some) human flourishing. What if we were to begin to think about planetary technologies for the benefit of life other than human life? What would technologies promoting the thriving of rivers, forests, or hybrid identities that transgress human categories of knowing the world look like vis-à-vis technologies that understand the rest of the planetary community as a standing reserve for human flourishing? Here "boundary" is not seen as ontological, but rather aesthetic: as human beings, we must begin to think about what types of futures we want to help cocreate. In other words there is nothing given about a particular boundary or identity. These are not fixed by nature or universal foundations but rather are always in flux, reiterated performances of various natural-cultural becomings. As philosopher of science Karen Barad notes: "Matter is produced and productive, generated and generative. Matter is agentive, not a fixed essence or property of things. Matter is differentiating, and which differences come to matter, matter in the iterative production of different differences."[5]

The "boundary" reached by the current regime of globalization is that it is leading to the erosion of space for multiple planetary identities. Granted, such globalization can continue for a long time, and the planetary community will survive beyond human truth regimes, but this book asks the more interesting question of what types of meaning-making practices we might co-construct to offer alternative planetary regimes to that of globalization. Thus the boundaries we have reached under the current regime of globalization are represented as obstacles to thinking past such issues as identity politics, human exceptionalism, rapid species extinction, and global climate change. These are ecosocial boundaries to be sure, and they call for planetary alternatives. To articulate such planetary regimes, I must begin here with stories from multiple perspectives. Such a multiperspectival approach will always be limited by those perspectives that go into my own processes of becoming, but nonetheless articulating the emerging planetary way of thinking can only be done from within, through narrating my own experiences of the effects of the regime.

On the one hand, the book's title, *Religion and Ecology: Developing a Planetary Ethic* seems to be about everything. I might as well just begin and end with the ancient chant: *Om*. On the other hand, what I do in the following pages feels like a description of a certain experience of the world at the beginning of the twenty-first century: an experience of a white, gay, male academician living in the one-fifth world and trying to figure out how to think ethically and critically from this hybrid position about how we ought to live vis-à-vis the rest of the world (both human and non).[6] These humble and honest beginnings won't quite do in a post-Enlightenment world where knowledge is no longer about compelling/forcing the other into submitting to the one truth, but rather is more about persuading planetary others toward viable ways of future becomings. I must, instead, persuade you that this *Gedanken* experiment is worth it by offering you several proverbial bones, one of which you will hopefully chase after.[7]

It wasn't until I was about two-thirds of the way through my first trip to India under the auspices of studying Jainism that a way to begin this book dawned on me. As I sat at a random hotel restaurant in Dharamsala, at the foot of the Himalayas, overlooking the exile residence and temple of His Holiness the Dalai Lama, while eating a Tibetan-Indian fusion meal and pondering whether I should get the sapphire ring I just looked at for my fiancé (who is first generation born in the United States from a Pakistani-Muslim family), an idea for how to begin came to me. All I need to do to introduce the topic of this book on religion, science, and globalization—all I can do—is tell a few stories from my own planetary context about what planetary identities and ethics might mean. These stories, no doubt, will not resonate with every reader, but no words do or can. The days when a single narrative (such as Western civilization) can pretend to capture the attention and experiences of all are long gone, in fact they never were. However, there ought to be a few "common grounds" of experience on which readers can stand.[8] Paying particular attention to issues of hybridity, which is the basis for understanding planetary identities, I decided to break the introduction into three different (though hybrid) beginning narratives. The book can be read from any one of these beginning points, and, read together, my hope is that they provide a common ground for a more robust analysis of life at the beginning of the twenty-first century. The hybrid narrated beginnings, then, will use the three major lenses through which the book is written to focus our attention on specific components of planetary becoming: the scientific story (roughly a new ontology), the religious story (roughly a new epistemology), and a planetary story (roughly a new grounds for identity and ethical concern). Let me begin here, then, with the scientific story.

A PLANETARY STORY OF SCIENCE

> The cyborg is a condensed image of both imagination and material reality, the two joined centers structuring any possibility of historical transformation.
> DONNA HARAWAY, *Simians, Cyborgs and Women*

As feminist philosopher of science Donna Haraway's infamous essay on "cyborg ontology" suggests, we are always and already cyborgs: our identities are material-ideal. Further, the world around us is always already a combination of imagination and material reality. Technologies are, indeed, a part of the rest of the natural world and shape the evolutionary process of planetary becoming. Many of my students and many people I speak with in general try to refute this common ground of living in the twenty-first century, but a few questions usually suffice in persuading others to the relevance of my point. How many people wear glasses? How many people would not be alive without the intervention of some form of Western, modern science during childhood?

How many people rely on pharmaceuticals for their day-to-day life? How many people rely on computers, the Internet, digital and wireless technologies to perform on a daily basis? This is just the tip of the cyborg iceberg.

The wider understanding comes with the realization that human beings, as meaning-making creatures, are also (as philosopher Martin Heidegger pointed out) technological beings.[9] As such, humans have never existed without some form of *techne*. In fact, if we widen our understanding of technology to include its etymology in Ancient Greece, *techne* (art/knowledge) is not just tool use (which is a later distinction made between the mechanical and ideal understandings of *techne*), but something used to order reality. Technology and concepts are equivalent: they shape realities and bodies. Both religion and science, and in fact all ways of knowing, are forms of technology. They shape our worlds, our bodies, and the worlds and bodies of others. In other words, ideas and materials are equally real. A thought can be just as strong as a punch in the face, and neither could exist without the other. Whether "ocean," "tree," "dog," or "table," concepts help to shape our realities into specific ways of becoming. And, as humans, we are born into technologies that shape our own becoming in ways over which we have no control. Unlike the idea found in complete constructivism, this book argues for coconstruction between humans and earth others, history and planetary evolution, and multiple understandings of planetary futures.

Though I begin here with the story of what is commonly referred to as modern science, it is only because I must begin somewhere. Further, this beginning gives the biohistorical contexts for exploring the differences between the technologies of globalization and that of planetarity. As the postcolonial literary critic and translator of Jacques Derrida (the father of deconstructionism) Gayatri Spivak notes, globalization is about domination of difference in the name of the same and planetarity is about connecting through differences. "If we imagine ourselves as planetary subjects rather than global agents, planetary creatures rather than global entities, alterity remains underived from us; it is not our dialectical negation; it contains us as much as it flings us away. And thus to think of it is already to transgress."[10]

It can be argued that the story of modern science itself is one of multiple histories, cultures, geographies, and beginnings; in other words, it is a story of connecting common grounds through differences rather than in spite of them. Modern science, though it often denies its own narrative underpinnings is, in a very real sense, planetary through and through. It is not produced without contributions from Indic mathematics, Ancient Chinese chemistry, Greek natural philosophy, Jewish and Christian monotheisms, developments in optics and medicine during the Golden Age of Islam, and contributions from literally all over the world of indigenous knowledge about medicinal plants and herbs. Yet this planetary story of modern science is "backgrounded" (ignored or made invisible) with the very adjective used to describe it, *Western*.[11] Chapter 1 begins to unwind this colonizing narrative of science that pits science against religion, humans against the rest of the natural world, and Western culture as the privileged place from which modern science is imposed over the face of the globe. I argue that religion and science (and humans and nature for that matter) are always already implied in each other from a planetary perspective. The myth of the Renaissance, the Enlightenment, and its "dark side" serve to cover over these planetary contributions, but only at the peril of excluding many others.[12] From this planetary perspective, the modern science touted as Western is yet another traditional ecological knowledge (TEK): that is, it is located in a specific historical and cultural trajectory. Often modern science is juxtaposed with traditional ecological knowledge in an attempt to delegitimize TEK and suggest that modern science is objective and universal. This book challenges these assumptions, and chapter 2 then begins to unravel the way in which modern science (as a traditional ecological knowledge) has coded nature as dead matter.

As I argue, the modern sciences understood as Western are also TEKs, coming from specific historical narratives that write nature as dead and humans as somehow exceptional to the rest of the natural world, and yet this dead nature is uniform with natural laws that were at first assumed to be created by a good creator god (though this part of Western science's story is conveniently elided). After uncovering some of the religious values smuggled into Western modern science in chapter 1, chapter 2 begins to challenge this stable, foundational, naturalized understanding of nature. In other words, the very construction of modern science has been so covered over that it is assumed to be universal, natural, or the way the world is. Here I argue that, whereas modern science finds resonance with the metaphysics of substance and monotheism (Aristotle, Newton, Christianity, Judaism, and Islam, for example), contemporary sciences, some call them postmodern, have found much more metaphorical and historical narrative support within Eastern traditions such as Buddhism, Hinduism, Confucianism, and Jainism. In other words, these (comparatively) nonsubstantive (or phenomenal-based) religious understandings of nature in part influence a science that takes these meaning-making practices more seriously and begin to reread and organize the world differently.[13] Even within the multiple stories that get molded into the false idea of a single, Western history, there are multiple trajectories that write nature and humans as processive, nonsubstantive, and thus understand nature as nonfoundational. This chapter will look specifically at the stories of Giordano Bruno, Spinoza, and the early emergentists as multiple starting points for understanding nature as *natura naturans* ("nature naturing" or "nature as process").[14] Such a view of nature is indeed the "death of nature" understood as something stable, out there, toward which we can conform or which we can preserve or restore. This is a somewhat more nuanced understanding of the "death of nature" from that which ecofeminist, environmental historian Carolyn Merchant describes in her book, *The Death of Nature*, and this view brings with it new technologies for human relations and human-earth relations.[15] In preventing nature from becoming a justification for conceptual binding, queer theory will help challenge stable notions of nature and identity that serve to keep humans locked into local identities: of self/other, human/non, male/female, straight/gay, etc. At heart, queer theory suggests that concepts are permeable and identities are always being coconstructed. Rather than something that becomes an apolitical policing mechanism of what counts as natural or unnatural, "nature" (in dialogue with queer theory) becomes more like an evolutionary rainbow of possible planetary becomings for which we can take responsibility.[16] These new nonequilibrium, queer, nonsubstantive sciences will require of us new ways of making meaning of the world as well, since meaning is no longer (and from this perspective never was) "out there." Chapters 3 and 4 begin this exploration from a perspective of a planetary story of religion and meaning making. It is to this story that I now turn.

A Planetary Story of Religion

> Indeed, the environmental crisis calls the religions of the world to respond by finding their voice within the larger Earth community. In so doing, the religions are now entering their ecological phase and finding their planetary expression.
> MARY EVELYN TUCKER, *Worldly Wonder*

As one of the leading figures of religion and ecology and cofounder of the Forum on Religion and Ecology, Mary Evelyn Tucker, suggests, the planetarity of religions is, in part, about religious reflection that recognizes its biological and historical embededdness in evolutionary and ecosystemic structures. Chapter 3 explores what this means for meaning-making processes in depth and suggests that religions are ecosocial constructions across multiple species, over multiple generations, and over multiple histories. More important, this recognition, I argue, in no way undermines the reality of religions or their importance. If we are cyborgs, hybrids of material-ideas-technology, and if cultures and religions are emergent realities of nature naturing, then ideas, reason, numbers, meanings, and values are realities equally as real as the material worlds of which we are a part.

In a recent course on methods in religious studies I argued that religions are natural-cultural projections. Students that leaned toward the atheist side of the meaning-making spectrum really liked this because they thought I was arguing that religions were somehow not real, but merely a feature of human psychology.[17] Students that leaned toward the confessional/theistic side of the meaning-making spectrum hated it because they assumed the same thing. The students were all wrong in this assumption. From an emergent, planetary perspective, religions are real in the same way that other things we cannot see are real: numbers, ideas, language, and imaginations, for example. What I ask my students to think about and now ask the reader to think about is any object around you, such as a chair, a table, or a house. These are perfect manifestations of ideal-materials. They do not exist without the reality of what one would describe in common parlance as ideas and matter. I argue the same is true of meaning-making practices. However, I am not arguing along the dualistic (and Idealistic) lines of Platonic Forms that matter fills; rather, I am arguing that form and matter are always already together and both are evolving and emergent phenomena. What we call matter and forms are akin to the difference between particles and waves. They are not different things, but different ways of describing the same evolving, emergent realities of which we are a part. From this planetary perspective, ideas (meaning/religion) and matter (nature/science) are both evolving and emergent and always already exist together. As such, what is directly opposed to this meaning-making process is certainty and dogma: anything that closes off the open becoming of identities.

One of the features of emergent phenomena is that they are open toward the future and toward the past.[18] They are, in a sense, living in an agnostic present. This is the same with our meaning-making practices, and many different religious traditions and philosophies have long histories of unknowing: deconstructionism, negative theologies, Jain relativity, and the concept of *neti-neti* in Hinduism (not this, not that) are good examples. I suspect that these iconoclastic edges of religions, if you will, and their mystic counterparts are a feature of our embededdness within a continuously evolving present. In other words, our five (and some might argue six) senses, even aided by technologies, can only sense so far into the past and future until our knowing shades off into mystery. Even in science we can't get behind the singularity theorized prior to the big bang or beyond the expanding edges of the universe. Some choose to fill these mysterious spaces with the certainty of a robust atheism (there is nothing!) or a robust theism (God or Ultimate Reality!), but both these positions claim too much. They seek certainty where none can be found, and certainty is the end of any faith and thus genuine hope for the future. Many forget that the opposite of faith is certainty, not doubt. Doubting is a part of what it means to dialogue with human and earth others. If one is unable to doubt, one can only have a monologue. More important, the type of ultimate, metaphysical closure that a monologue takes manifests itself in smaller everyday closures that can be violent toward human and earth others: fundamentalist beliefs, dogmas, bigotry, and just about all "isms" can probably be linked to some sort of ultimate certainty.

Planetary religions (or meaning-making practices) will be open toward the evolving, emergent contexts from which they arise and to which they return and shape human and earth others. We can think of meaning-making practices as multiple "lines of flight" that seek to persuade and fashion realities into certain ways of becoming.[19] The French postmodern thinkers Gilles Deleuze and Felix Guattari suggest that creative thought is like a line of flight that tries to move life into ever new and creative ways of becoming. As such, these lines of flight can shift with shifts in our awareness of the ways in which our practices affect other, becoming bodies and our own identities. Such shifts are the subject of chapters 4 and 5.

We are ecoreligious, biohistorical subjects through and through. This is perhaps the silver lining of the problems of global climate change and the ill effects of globalization, not to mention the information about the worlds we inhabit coming from those sciences we call Western. If nothing more, because of these social and ecological problems, we recognize that our identities, whatever else they may be, are not located in some transcendent place far away; they emerge out of the planetary processes from which all other life emerges. Furthermore, and more important, our creaturely meanings return to affect our own and other becoming bodies. Perhaps, for instance, living as if we were made in the image of God or somehow over and above the rest of the planetary community has led to the very sorts of relationships between humans and earth others that have resulted in the sixth great extinction period and global climate change. The point is not that these things are unnatural, but rather the point is to ask if this is the kind of planetary community we want to help cocreate. As meaning-making creatures like all other living creatures we are response-able (able to respond to others around us). Our unique characteristic as a *Homo sapiens sapiens* (though not exceptional) is that our response-ableness has evolved into responsibility. We are responsible (to varying degrees) for how our meaning-making regimes and our truth regimes affect the lives and bodies of others. One may argue that such an understanding or collapse of truth and meaning subordinates truth to meaning, but I would argue that this has always been the case. The need and desire to transcend reality (as is found with anthropocentrism and human exceptionalism) is in some way the result or desire for justice and goodness to win out in the end. In other words, the whole idea of truth as a transcendent reality (whether attained through religious revelation or objective scientific observation) is about the desire for meaning: whether that is a just and good creation or an understandable, predictable, and manipulatable world. If there is no away realm from which we will be judged or that can save us from our ignorance, the planetary community is then recognized as the beginning and ending place for ethical judgment. There is no safe zone away from the hard work of coconstructing our lives, taking responsibility for our coconstructions, and beginning again to think toward alternative futures. As such, we must begin to account for the transcultural and planetary contexts of our meaning-making practices.

I have argued elsewhere that it was much easier to perform a Christian or Jewish or Buddhist or any other meaning-making practice as the only practice, and in fact to project it as the truth of the world, when these meaning-making practices were relatively isolated (geographically and historically) from one another.[20] However, colonization and globalization have brought about greater awareness of our hybrid cultural-religious identities and practices. Our Lady of Guadalupe, American Christians who do yoga, Indonesians who are Muslim but still practice indigenous medicine and perform a Hindu ritual here and there, Hindus that pray even at times to the god of the Christian tradition, and even atheists that meditate or find Eastern philosophies helpful are all various manifestations of the transcultural religious identities I am trying to articulate. Some are forced and others are "chosen," but, regardless, these crossings highlight the fact that meaning-making practices are embodied performances affecting the way we dress, what we eat, our music choices, and other daily rituals performed.

Once multiple practices are readily available, rituals are revealed for their practical use: mediation leads to greater clarity rather than being the only spiritual practice available and thus *the* way; the Bible becomes one way of relating life to the divine rather than *the* divine word; etc. These performances of meanings (because we can recognize them as such without disparaging them as somehow less meaningful) can both bring humans and earth others together in delightful celebrations of life and lead to some of the planets greatest miseries. From a planetary perspective, we must respond to how our meanings materialize in the world in better and worse ways. The focus of chapters 5, 6, and 7 is precisely on the ethical harvest from understanding religion, nature, and science in these response-able ways. In these final chapters, the ethical take-home of thinking ourselves as planetary creatures begins to emerge.

Planetary Stories: Identities and Ethics Beyond Exceptionalism

> Despite their real and important differences . . . Western theological and philosophical traditions [all assume that]: human beings are not fully at home in the world. What is really important, really *human* about humanness is the extent to which it transcends its physical and social existence.
> ANNA PETERSON, *Being Human*

If we understand humans and our meaning-making practices as emerging out of the processes of planetary evolution, and scientists as speaking on behalf of multiple perspectives of a nature that is always in process, then the usual categories we use to seal our identities off from human and earth others fall to pieces. In other words, our identity is no longer based on an essence that one finds in a transcendent ideal location or in nature. Both are processes on the move, and thus even the laws we think of as "natural" are emergent and evolving. Likewise, the interpretation of our religions and the realm of meaning making is always adapting and evolving. Again, our identities are a part of this process: through and through we are planetary beings. This has immediate implications for both ethics based on foundational categories and those based on identity politics of liberation. Neither quite holds up under the conditions of planetarity. The more ethics-focused section of the book, then, begins with a discussion of foundational-based identities in chapter 4.

Identity politics are directly correlated with how we understand religion and science and thus shape directly our attitudes toward the rest of the natural world. In a world where sex, gender, race, and sexuality have been used to seal people off from others (including nonhuman others), queer theory becomes a valuable source for thinking about identities from planetary perspectives. Queer theory thus also becomes a source for critiquing the economy of relations: i.e., how we relate to one another and our rules for use and nonuse of others (human and non). Chapter 5 will look at the case of Caster Semeyana (the South African Olympic Runner accused of being a male and not a female), the Waria (third gender of Indonesia) and the growing pro-intersexed rights movement in the medical profession (as many as one in a thousand human babies are born intersexed) to argue that sex, gender, race, and sexuality are coconstructed with the more than human world, and, when opened up to this processive, planetary context, they can become radical locations for resisting the capitalistic economy and corresponding capitalistic politics of love. Through these stories and the identities that are narrated by them, one can begin to see the freedom involved in opening one's thinking and acting toward planetary becoming, which is the beginning point for understanding the need for an environmental ethic of movement (chapter 6) and thinking past human exceptionalism or the possibility of a posthuman future (chapter 7).

If our identities, natures, and meaning-making practices and values are always already on the move within this planetary context, then the ethics for how we deal with scientific technologies, environmental issues, and bioethics will no longer make sense by referring to what is natural or god given. Rather than an environmental ethic of place, which smuggles in some heterosexist assumptions about the importance of fidelity to a single place and ideas about pure nature toward which we can return, we will need a bit of polyamory toward places.[21] We need, in other words to learn to love multiple places in order to understand their planetary connections. Drawing from previous chapters, I will argue that our ethical assertions always already begin on slippery slopes (no foundationalism: we begin from uncertain places), but nonetheless from common grounds. These common grounds include the planet, the universe, laws of gravity, animality, and humanity, for instance. They also include the need for making meaning. However, these are not static common grounds: they can shift, shake, erode, and change. Starting from such common grounds and moving toward uncharted territories will trip up the desire to impose ethical and value sameness over the face of the globe and suggest a polytheistic (or at least polydox) planetary nomadism.[22] Such a nomadism adopts local gods and loves many places, yet maintains identity through the unique narrative process itself: harkening back to the quote opening this introduction, no one has the same set of ecosocial, historical, and evolutionary influences past, present, and future. Thus unique identity is through and through a conarration of all of these elements. As such, our understandings of agency must be rethought.

Planetary environmental ethics will inevitably be contextual and multiple, yet I argue they will have the commonality of challenging agency, understandings of integrity and hypocrisy, and, ultimately, understandings of love. In other words, the idea of individual human agency, the charge to be integral at the peril of ethical hypocrisy, and the forms of love that identify with mere hospitality and notions of peacekeeping vis-à-vis the other are precisely the sources of social and ecological ills brought on by a monotheistic mindset that keeps the individual locked into identity politics and disables planetary becomings except for creating the world in the image of the same. We need to: recognize our coagency, realize that we will be ethical hypocrites arguing for new ways of being while still living in the globalized capitalist consumer society, and embrace iconoclastic love that argues with difference in order to not gloss over those differences in the name of the liberal notion of equality (or the idea that we are all really the same inside). We must not only do this with human others, but with earth others as well, which is the subject of the concluding chapter.

As planetary, nomadic subjects with multiple possibilities for future becoming, we must also challenge the very essentialism of being human. We must seriously think about what it means to become plant, mineral, and animal with others in the planetary community present and future.[23] On the other side of the spectrum, as we begin to understand humans and human thought and technologies as nature naturing, we must also embrace our hybrid cyborg natures. As Philip Hefner notes, "'Human Becoming' expresses the idea that we are always in process, we are a becoming, and being human means that the journey is the reality—there may well be no final destination."[24]

One such metaphor in service of this thought exercise is Timothy Morton's notion of hyperobjectivity and hypersubjectivity.[25] In other words, the places where we make identity boundaries could be at the individual level of rock, river, or human, for example, or we could think of ecosystems as a hyperobjects/subjects of which we are parts or global climate change as a hyperobject/subject of which we are parts. Indeed the processes described here as globalization and planetarity could also be thought of as hyperobject subjects. Following through with some implications of this thought experiment, and the challenges to religion and science throughout the text, I will argue for developing planetary sciences and technologies. Such sciences and technologies would recognize that science and meaning making is not just *for* human beings, but *for* the becoming hyperobject subject that is the planetary community. We can then begin to consider a science and technology for trees, rivers, polar bears, and oceans (for instance), rather than merely considering how technologies will curve the earth toward human needs and thereby continue to reify the false assumption that humans are not evolving with the rest of the planetary community. Through imagining such planetary technologies for becoming, we open many possibilities for future becoming beyond the logic of sameness imposed by forces of globalization. Thus this book in the end hopes to offer sites of planetary resistance against the powerful forces imposing conceptual violence upon identities and relationships between earth others.

In an effort to avoid conceptual violence within these pages, this book will be dealing with a lot of terms and concepts that have many interpretations and histories. In order to operationally define some of these terms, I have included a glossary at the end of this book. The reader should not assume that the definition s/he finds in that glossary is the only (or even the best) definition or interpretation of a term or concept, rather it is meant to describe how the term or concept functions within this book. The glossary, then, is a good place to begin the process of understanding difficult terms, but is by no means the last word.

1 RELIGION AND SCIENCE IN DIALOGUE

> What formerly happened with the Stoics still happens today, too, as soon as any philosophy begins to believe in itself. It always creates the world in its own image; it cannot do otherwise. Philosophy is this tyrannical drive itself, the most spiritual will to power, to the "creation of the world," to the *causa prima*.
> FRIEDRICH NIETZSCHE, *Beyond Good and Evil*

One of the most crucial insights of the last four hundred years, I would argue, is the one that Nietzsche articulates in the opening quote of this chapter. Philosophies and religions, science and matter are but different ways of trying to account for the worlds we inhabit: neither is transcendent to the other, but rather both exist always already together. There is no origin, no transcendent point from which either an idealist or materialist account of the world can recapitulate reality because eventually they must account for their own accounting and the ouroboros of thinking and being eats its own tail. Many of the thinkers I will deal with in this book are struggling with this same problem: how do we know about the world without transcendence if we ourselves and our very knowing are always already shaping the worlds we seek to know? In various ways, peoples have argued, there is no outside, no pure origin, no telos (or ultimate goal of the universe), and no thing in itself transcendent to the process of ongoing nature. And, still, we must begin to think, make judgments, and act in the world according to this knowledge that is now no longer "The Way Things Are," but rather, at best, a multiperspectival, contextual account. Here in this chapter I want to begin telling the story of the relationship of Western science and religion from this perspective of always already being intertwined with each other. Many accounts exist narrating the dualistic split between the two and the effects of "the secular" that results from that split.[1] I will draw on these accounts, but what I want to do here in these next few pages is begin to narrate a planetary story of religion and science and their relationship to nature. As mentioned in the introduction, "planetary" is meant to challenge the monological logic of globalization. Such monological logic would have science and religion as two separate bodies of knowledge in some sort of opposition, conflict, or perhaps even as different, equally valid ways of knowing the world or as two separate ways of knowing that will eventually converge on a single truth about reality.[2] However, this book begins from the assumption that these two different methods for understanding the world are always already together. I am not trying to reenchant nature or bring religion and science together or even bridge the ontological gap wedged by the Cartesian Cogito, but rather I am assuming that there is no pure secular space, that religion and science are always already together, and that the gap does not exist. From this perspective, we might begin to see a history of religion and science that does not reinforce the split, but mends the wounds that have so long been reopened by the technologies of a secular, materialistic scientism, on the one hand, and a private, immaterial religiosity, on the other.

The main problem with the dialectic of the Enlightenment, as philosophers Theodor Adorno and Max Horkheimer intimated, is that the very foundations for the ideal and material do not exist.[3] Thinking about biology and history, religion and culture, ideas and matter as always already together will be a different way of thinking than most of us in the contemporary one-fifth world are accustomed to, so I begin here with a brief description of a nuanced version of Agrippa's trilemma (or the trilemma that results from attempts to make knowledge claims), then I move on to define the loci of concern for this chapter—religion, science, and nature—and finally toward a story of their historical relationship of mutual interaction. From this place, science and meaning-making practices help to cocreate planetary communities for which we can be responsible. Furthermore, they have long been involved in coconstructing the very planetary identities that are at the heart of this book.

A New Version of Agrippa's Trilemma: Toward Planetary Knowing

In epistemology, or the study of knowledge, the question of how we justify any knowledge claim is often captured in what is known as Agrippa's trilemma. In this trilemma the question of how to justify a knowledge claim is answered in one of three ways. The first two are representative of the type of dominology that globalization of sameness exemplifies. If we propose a planetary way of knowing over that of a global, then we must find a different way out of this trilemma than those that have been suggested by foundationalism and circularity, both of which end up in a form of solipsism. In a circular argument, argument A is based on B, B is based on C, and C is based on A again. Strangely, though credited with being a foundationalist, Descartes's understanding of the cogito (the "thinking thing") and the existence of God makes this type of circular argument. The cogito is posited and then becomes the foundation for a good creator God, which is then justified by the cogito. The circular argument, then, is beholden to solipsism. Solipsism keeps us locked into our own knowledge assumptions and identities through enabling us to background our relations with multiple earth others. In this case our identities and knowledge projections are like a boomerang that come back to affirm our original assumptions in a circular manner. No new information challenges the construction; the "I" or cogito can remain untouched because it is founded in the transcendent Creator, and the Creator is subsequently founded in the assertion of the "I." In this manner, the I-God circle will construct all it senses in its own image: creating the world in the image of sameness.

Another option in the trilemma is foundationalism. In foundationalism a claim is posited sui generis or a priori, needing no other conditions or causes on its own to exist. Again, Descartes's "Clear and Distinct Ideas" would be an example. Aristotle's "unmoved mover," which sets the whole cosmos in motion but does not itself move, is also an example. Another example would be the idea of *creatio ex nihilo* or "creation out of nothing."[4] The purpose of a foundational claim is to stop the slippage of knowledge by making an assertion that cannot be questioned. From such a smooth place or foundation, knowledge claims force all reality to comply with their own assumptions. In other words, the foundational option leads to universalisms. Foundationalism operates (most often) by digging down to what is perceived to be a base of reality: whether material (as in scientific materialism) or ideal (as in the creation of the world according to divine laws by a good God). Think of a foundational claim as a river dam, which floods a valley creating a lake and controls the flow of water from one side to the next. These dams cover over the details and create realities (a lake) according to their own logic. Or, if in your childhood you ever enjoyed a Play-Doh Fun Factory, think of a foundational claim as one of the attachments that you press the Play-Doh through, thereby making it conform to a certain shape. Our foundational claims work in a similar way. Furthermore, these foundational claims also end up in solipsism, as the whole of reality is distorted to fit into the central foundation.

The logocentrism critiqued by Derrida and so many deconstructionists operates precisely in this foundational manner.[5] In logocentrism all reality is made to fit into the confines of the logos—word, concept, telos, salvation narrative, or reason. Let's take, for example, the narrative of Christianity. If one believes that the world is created by God "in the beginning" and is moving toward a salvific point in the future, then all life must fit into this (Christian) salvation narrative. This is a distortion of the realities of multiple human and earth others. This same sort of foundationalism gets taken up into understandings of scientific progress (replace revealed creation with natural laws and revelation with reason) and in ideas of economic development (replace fallen creation with primitive, and salvation with civilization). In fact, as Gayatri Spivak intimates with her juxtaposition of planetarity and globalization, globalization itself operates along the lines of this same type of foundational logic.[6]

The third option in the traditional trilemma is infinite regress. From this perspective there is no stop to the slippage of knowledge claims. All claims are "ultimately" unjustified and can be questioned ad infinitum. Socrates knew this well, which is why his pesky Socratic method annoyed so many Athenians during his day.[7] Anarchism, dadaism, and deconstructionism, among other camps of thought, move toward this type of understanding of knowledge (from different contexts and toward different purposes, of course). Though the least popular option in epistemology, I argue that this is the only option that breaks our thinking and concepts open onto the contexts and contours of the worlds in which we live. It is, then, the only option that breaks us out of dominology. In other words, both circularity and foundationalism secure knowledge from changing contexts: they cut our knowledge claims off from reality and force reality to fit in to the claims made about reality. Infinite regress, on the other hand, highlights the fact that we are all contextual, perspectival, embodied, changing creatures and that our knowledge claims can thus never be justified in foundations or circular arguments; rather they are always made on shifting grounds.[8] As Mary Jane Rubenstein notes, "The point is that there is no point: no end to the uncertainty and no solid ground on which to construct a doctrine of knowledge or, for that matter, of being or its truth."[9]

In later chapters of this book, I argue that this same trilemma holds for foundations of identity markers such as gender, sex, and sexuality (chapter 4) and that the infinite regress model offers some contextual, open, and evolving models for gender, sex, and sexuality. It is only with such a model, a model that can be described as queer, that the violence caused by identity essentialisms gives way to creative, multipossible becomings of planetary identities. Further, a practice of such infinite regress in identity construction also opens us onto the rest of the natural world (chapter 6) and toward the possibility of evolving past our own humanity (chapter 7). Queer theory is at least, though much more than, a deconstructive or iconoclastic attempt to break down conceptual binaries and break apart the process by which concepts constrain and choke identities and life so that multiple identities and lives can become and exist.

However, before going into these issues of identities and ethics, we must spend the next few chapters closely examining the two seemingly stable sources for any knowledge claim (whether justified by foundation, circular argument, or applying the method of infinite regress), variously described as nature/science/material *and* culture/religion/ideal. Both circularity and foundationalism can ignore one side of the way in which humans know about reality. In other words, it is possible from a foundationalist or circular perspective to be a complete idealist or a complete materialist (though only, as I will argue, in theory, since both are always already together). One unhelpful solution that brings the ideal and material worlds together found in foundationalism and circularity is that of dualism. Foundational dualism often claims that one side of the spectrum—material or ideal—is really real. Circular reasoning forms dualisms through using the ideal to justify the material or vice versa. Only infinite regress takes both sides together as equal co-constructors of reality. As Rosemary Radford Ruether notes, we need both Gaia (material) and God (ideal) as human beings in the world. "Both of these voices, of God and of Gaia, are our own voices. We need to claim them as our own, not in the sense that there is 'nothing' out there, but in the sense that what is 'out there' can only be experienced by us through the lenses of human existence."[10] We can't seem to escape the fact that we are always already natural-cultural or biohistorical creatures.[11] For this reason, the fields of "religion and science" and "religion and ecology/nature" become important sites for thinking about what it means to be humans within a planetary context. Let me begin the examination of these two different approaches to knowing the worlds around us—science and religion—with a few, brief operational definitions of what I mean by these terms. The following three terms have multiple meanings, and volumes are written on each, so the best I can do here is describe what I mean by religion, science, and nature within the context of this book and within the context of constructing planetary identities.

Defining the Terms: Religion, Science, and Nature in a Planetary Context

Religion, again at its etymological root, is about rebinding and rereading. It is about putting together a meaningful world, given the information we have within our specific (evolving) contexts. To stick with the trope I have been using, religion is about meaning making. Following the (somewhat modified) argument of Martin Heidegger, among many others, religion is then one of the ingredients of worlding or enframing.[12] As such, humans are, if anything else, meaning-making creatures. If birds fly, dolphins swim, and dogs bark, humans make meaning and are made by this meaning. I am not arguing that humans construct meaning out of nothing and create worlds or that language creates reality. This would be a monological process of imposing meaning onto the entire planet. Rather, humans are born into worlds of coconstructed meaning. These meanings are coconstructed by biohistories of humans living within a larger earth community. Other animals, plants, and the rest of the natural world help humans make this meaning; just as a bird would not fly without gravity and the dynamics of air and a dolphin would not swim without the history of evolution leading to its specialized form and the ocean water in which to swim, so humans do not make meaning without the histories, languages, genetic evolution, stable climate, and surrounding earth communities that make meaning possible.

To say that humans are meaning-making creatures is not to suggest that all humans are religious in the sense of following a traditional world religion, but rather that we all—atheist, theist, or agnostic—make meanings out of our lives that place us into a wider context of human-other and nonhuman-other relationships. These meaning-making practices matter and matter to our bodies and the worlds around us: they shape our ideas about gender, humans, the more than human world and they shape our cultural institutions—legal, economic, and political. These ideas and institutions then shape the many bodies that make up our worlds. Following Durkheim, among many others, religions function as binding forces or glue in our societies and daily lives.[13] Religion also functions to help us cope with existential matters such as life-transitions, illness, and death. As Thomas Tweed notes, religions provide us with both crossings (navigating sticky existential transitions) and dwellings (making sense out of the newfound worlds in which we find ourselves).[14] Religion, then, can also be understood as any system that organizes our life into a meaningful daily existence. Consumerism, the free-market economy, environmentalism, and other such systems can be analyzed as meaning-making practices, and in this sense they are religious.[15] To ask the further question about "why" these meaning-making practices have evolved can lead to a lot of interesting thought experiments. However, these must be understood as speculative and imaginative. There is no reason why life has evolved into birdsong and meaning-making creatures, or so I will argue, but rather this is the gift of life we have been given. It is the mystery of life, something that evokes wonder and awe, and it begs for what I will describe later as a viable, planetary agnosticism.

Finally, when I use *religion* here, I can never stress enough that religion is always already a Western construct in history and formation (just as are Western science and nature, which will be discussed below). Confucianism and Daoism may be more like philosophical systems than religions; Hinduism is more like a conglomeration of different local practices and beliefs that are inseparable from the larger culture of India; and many indigenous traditions are more like lifeways or socioscapes than religions. In Muslim nations such as Indonesia and countries such as India, one notices that daily life is much more affected by what Westerners might think of as the religious. Whether in dress, public call to prayers, daily devotionals to deities, etc., the secular as a separate space really doesn't exist as it does in the West. As I will argue later in this chapter, it takes a split between humans and nature or religious thought and natural philosophy such as occurred in the history of the West to get such an understanding of "religion" as a separate realm from the rest of public life and culture. Hence the globalization of the idea of religion as a private matter and modern secular science as something universally available to all is itself rooted in the historical relationship between religion and science within Western history. This is one reason that I will argue in the next chapter that Western modern science is itself a traditional ecological knowledge that understands nature as dead/secular. As a result of this split, or "great divide," we must understand modern science too as Western in its construction, and it is to a definition of what I mean by this science that I now turn.

The root of science or *scientia* is *scire*, which originally meant to separate one thing from another. *Scientia* more commonly means knowledge or knowing. Taken together, we might creatively suggest that science implies a method of knowing through separation, taking apart, examining specific parts of the worlds around us. Science focuses on material, and energy flows within and between its objects of study. It includes examinations of the smallest levels of reality (quantum and subquantum physics) and the largest levels of reality (cosmology), and every level of examination separates itself from the other in such a way that integrative analyses are needed to bring the various levels back together.[16]

It is always important to remember as well that science also involves scientists. Scientists, as human beings, come from various social locations and bring their bodies and contexts to the data. Science, then, must also always be engaged in the critical reflection on how subject positions and contexts shape interpretations of data. Thus the practice of science includes reflections from philosophy of science. Such reflection, as Sandra Harding notes, takes account of the multiple subject positions in interpretations of data: this is the only path toward a "strong objectivity" that reveals multiperspectivalism.[17] In other words, a "strong" objectivity would include the subject-position of the observer rather than ignore that the observer's history; culture, gender, race, sexuality, and socioeconomic status (among other markers) all shape the ways in which he sees the world. Any scientific study of nature always already includes culture. Karen Barad, whose work I will discuss further in the next chapter, sees this observer effect as a largely ignored implication of Niels Bohr's understanding of quantum physics. Such observer effects mean, for one, that truth is cocreated and that we are cocreated by these truths through structuring phenomena.[18] From a planetary perspective, then, we can begin to see how science is multiperspectival. Rather than one universal science, we can begin to study a multitude of sciences that extend well beyond the Western natural sciences. One might think of a science tied to Indian cultures or Japanese cultures, for instance. In fact, Ayurveda sciences and more holistic sciences of Eastern cultures are becoming more and more popular in Western practices of health and healing. Further, I would argue, these Eastern traditions are very influential on the development of what David Ray Griffin calls "postmodern sciences."[19] In other words, these Eastern traditions, even though never completely separated from Western traditions, help shape the constructions and regimes of knowledge for understanding nature in so-called postmodern sciences (chapter 2). A planetary understanding of science will stretch our meanings of science even further to suggest that we need corresponding planetary technologies (chapter 7). Such technologies will be not merely about organizing and using nature *for* human ends, but for the benefit of the rest of the natural world as well.[20] We might think of a science and technology of and for the trees, rivers, birds, dolphins, and oceans, for instance. This will require challenges to our understandings of *nature*, the third term I want to operationally define here in this section.

Nature is my all-inclusive term. It includes humans, cultures, religions, ideas, imagination, atoms, ecosystems, the earth, the universe, and all other levels of reality. Nature is multiscalar and emergent. By multiscalar, I mean that it consists of multiple levels, none of which can be reduced to the other. Thus, as Kevin O'Brien notes, "The multi-scalar reality of biodiversity will be conserved only if we learn to think in such multiscalar ways and to be honest about the limits of our attention."[21] By emergent, I mean that nature is a process by which "new" (*natal*) levels emerge in the course of planetary and cosmic evolution.[22] Thus nature is a multiperspectival emergent process. Nature is the *natura naturans* of Spinoza (see chapter 2), but without the *naturata*.[23] Finally, it should be noted that there is no birth without death. The new is never *ex nihilo*, but emerges from the destruction of the moment—energy and materials—of that which comes before. As some scholars of the theory of emergence suggest (see chapter 2), nature is the ongoing creative-destructive process of life.[24]

Given these definitions of religion, science, and nature, how might we re-read ourselves back into the rest of the natural world in a way that is conducive toward an understanding of human beings as part of a planetary community? In the rest of this chapter, I argue that we have indeed been reading our identities through understandings of nature as defined through a dialogical interaction between religion and science all along. In fact, most Eastern and indigenous traditions, philosophies, and lifeways have only been forced into separating religion and science through processes of colonization and its grandchild, globalization. What is new is not the method of making meaning, but the contexts in which meanings are made: namely, globalization and global climate change. These new contexts are enabling us to recognize the coconstruction of these categories as globalization begins to challenge the clearly delimited epistemologies, histories, and cultures we once thought existed and global climate change challenges the idea of a nature in equilibrium.

In order to investigate the worldings resulting from the dialogical interaction of religion and science, I first examine the beginnings of the Western separation of religion and science in the Christianized Ptolemaic understanding of the cosmos. Further, I argue that the very separation forged during the scientific "revolution" (what Carolyn Merchant refers to as the death of nature) itself is the result of "interstitial" or "hybrid" identity formation of later understandings of religion and science.[25] As such, my claim will be that religious thinking always already contains science and scientific thinking always already contains religion. Through the dialogic interaction of religion and science, our worlds are made more or less meaningful. Further, it is precisely at this juncture in our histories—marked by globalization, pluralism, hybrid identities, and a changing planet—that the very foundations of our worldings are being challenged. In this challenge, I argue, science and religion in dialogue can help redefine the human as part of an emergent, planetary process. As always, in order to imagine forward into the future we must first imagine backward into the past, or else there is no present place from which to begin.

The Ptolemaic Cosmology, the Scientific "Revolution," and the Grounds for Globalization

> The new conceptual framework of the Scientific Revolution—mechanism—carried with it norms quite different from the norms of organicism. The new mechanical order . . . and its associated values of power and control . . . would mandate the death of nature.
> CAROLYN MERCHANT, *The Death of Nature*

What Merchant describes in her book, *The Death of Nature*, is the process by which a specific culture and civilization ("the West") came to understand a divide between humans-culture and the rest of the natural world. This gap or separation placed "civilized," white, and male humans as the source of all value and wrote nature and the rest of the natural world, including "uncivilized" humans, as dead or passive matter waiting to be made valuable through cultivation, extraction, and transformation toward the progress of Western culture. The contemporary process of globalization spreads this "great divide" to the rest of the planet. In theory, of course, the Universal Declaration of Human Rights has extended the active/valuable side to all human beings. And, to a great extent, this can be seen as one of the chief triumphs of the modern, Western liberal understanding of the self. However, and as I will argue further in chapters 4 and 5, such a universal declaration, precisely because it is not grounded in the performance of actual identities in practice, has not been able to stem the tide of violence toward humans who find themselves on the colonizing end of the process of globalization. In other words, an imposition of universal, individual human rights over the face of the globe fails on some level because it does not succeed in acknowledging the always already of embodied biohistorical existence. Seeking to create a special category for all humans as over and against the rest of the natural world simply does not recognize the variety of ways in which human beings navigate their always already natural-cultural identities. Furthermore, this idea of human rights is founded in a liberal humanism that assigns the individual way more agency and responsibility than perhaps is possible, or so I will argue in later chapters.

I am not suggesting that individual rights are a bad thing! However, I am suggesting that it is the legacy of this theoretical great divide, with its theoretical demarcations of what is and is not human/active/valued, that has led to many of the contemporary ecological and social ills associated with globalization. As Philip Goodchild argues, humanism and its corresponding human rights depend upon the mastery of nature through science, control of nature through technology, and liberation from nature through economics.[26] This divide, which makes humans exceptions to the rest of the natural world (the subject of chapter 7) has roots in an earlier form of exceptionalism and can also be described as the process by which the divine or sacred exits the world. It is only at this point of separation that one can begin to talk about distinct categories of religion and science. In fact this separation is foreign for most peoples until processes of globalization force this separation upon cultures, bodies, and identities around the globe. This separation is more aptly described as a transformation of spirit and divine revelation into reason and natural laws. In order to explain this split, I begin here by reminding the reader of the Ptolemaic/Aristotelian understanding of the cosmos through which early Christian theology developed.

As many readers may know, Aristotle's universe (which was further developed by Ptolemy and is also known as the Ptolemaic universe) narrates the earth as the center of the universe with the planets and sun revolving around it. (Just Google Ptolemaic universe and you will find multiple images of this worldview.) Beyond these concentric circles are the realm of the fixed stars and the realm of the Unmoved Mover that sets the universe in motion and keeps it moving by infusing the whole cosmos with a "world soul." This structure is important for the concept of teleology, since all things within this understanding of the cosmos move toward their own natures. Each human spirit is in part determined by stars (hence astrology), and our spirits long to return toward their home in the sphere of the Prime Mover.[27]

Christian thought is deeply tied to this view of cosmology. Biblical Scriptures were read and interpreted through this cosmos, and the theology developed was thus dependent upon this cosmos.[28] Let me provide you with one example: the Trinity. One of the hardest things to explain about Christianity in its earliest stages was how God came to earth in human form. This is because, for the Greek mind, movement and perfection were diametrically opposed. There is no way the Unmoved Mover could materialize and move about on the earth. Christian theology had to develop a way to describe how an impassable God—all powerful, eternal, unchanging, omnipresent, and all knowing—could move about on the planet: it did so through Trinitarian thinking in which the Father (the Unmoved Mover) stays put and is at the same time incarnated in Jesus Christ through the mediation of the Holy Spirit. Entering this conversation are all sorts of doctrinal quibbles about the nature of Jesus and Jesus's relationship to the Father: one substance with two forms, two forms, one substance, etc. The point is that religion (in this case Christianity) is always already involved in making meaning out of a world where science (in this case natural philosophy) has much to say about that world. The same can be said, quite arguably, about most cultures, but in most cultures the split between what is science and what is religion does not occur in such an absolute way, so the statement is not quite accurate. It is even anachronistic to tease out science (natural philosophy) and religion (early Christianity or early communities following the teachings of Jesus) before the rise of modern science, but I do so here for illustrative purposes. Christian theology developed for almost one thousand years with this Ptolemaic cosmological model.

The Dark Ages were actually not dark at all; rather they can also be described as the Golden Age of Islam when Islamic scholars took up Greek thought and made significant developments in mathematics, optics, medicine, and what would eventually become modern science. Thus, the Copernican and Galilean "revolution" is not possible without input from the Golden Age of Islam.[29] Most accounts of Western history leave out the influence of the Islamic world on modern science, not to mention the period of *convivincia* in southern Spain, during which Christians, Jews, and Muslims lived together in relative (though not perfect) peace and toleration, or the influences of China and India upon modern science. Through backgrounding this religiously and culturally mixed history, contemporary Western peoples are more easily able to make Islam the "dark other," that which is a threat to the "light of reason" and the goodness of civilization, and claim modern science as Western.

Witnessing to this problem is the recent uproar of people over the proposed building of the Cordoba Institute near the site of ground zero in New York City. (Though technically, "the ground zero mosque" is neither at ground zero nor a mosque!) Cordoba was precisely chosen as a name for this Muslim community center because it was one of the primary cities of the *convivencia* and thus provides an historical example of Muslims, Christians, and Jews living together and exchanging cultures and ideas. It reminds us that our religious identities are always already mixed together: there is no pure tradition. Similarly, there is no scientific revolution without the contributions the Islamic world made to modern science, not to mention the contributions of Jewish thinkers such as Maimonides, and, as noted, mathematic contributions from India and early chemistry from the Chinese. In this sense, what we call modern science is truly already planetary in construction. To cut off these planetary ties and claim it as Western is precisely the type of violence perpetuated by a global mentality toward "others." Such claims, as Walter Mignolo points out, not only background the contributions of others, but set "other cultures" and knowledges on the side of "traditional" knowledges in need of enlightenment if not civilizing.[30] In other words, claiming modern science for the West eventually becomes a tool of colonizing and cultivating human and earth others, which/who are seen as in need of being brought into the single history that moves from the creation story in Genesis to the Greeks to the Romans and then to Christian Europe.[31] Such a singular (salvation) history begins to crack and bend under the pressure of trying to assimilate so much difference into its own sameness; one such cracking and bending is what we now refer to as the scientific revolution.

Effectively what happens during the scientific revolution is that the Ptolemaic cosmos begins to buckle. Copernicus argued, and Galileo later confirmed (through optical technology that was made possible by contributions from Muslim scholars), that the earth is not at the center of the universe and that we actually revolve around the sun. This was shattering to the Christian theology that had placed human beings (again, Christian human beings) at the center of God's salvation history based upon the Ptolemaic model of the cosmos. Furthermore, the movement of the stars threw into question the idea of a fixed universe. What was going on here was not so much a conflict between religion and science but between science and science: between the Ptolemaic cosmology and the newly emerging Copernican one. At the same time, the Reformations in England and Germany were challenging religious hegemony. Thus, between the uncertainty in religious authority and the uncertainty in the new scientific cosmology, much doubt was cast on the Christian understanding of the world.

It is with Descartes's cogito and Newton's billiard ball model of atoms that the newly emerging mechanical cosmology gained its most persuasive definition. Descartes's cogito effectively places value within the individual "thinking thing." All other life is just dead matter and only receives its value through human use. Furthermore, the cogito separates mind and body in such a way that the cogito is as a "ghost in the machine" of the body. With Newton, nature comes to be understood as dead matter, consisting of billiard-ball-like atoms that are moved by external forces. The whole cosmos becomes mechanical. God's role is only important as its creator. This deism (though Newton himself was not a deist) becomes very important for modern science. It means that the world can be manipulated and controlled toward the human project of progress, which is slowly taken out of the hands of religion and priests and given over to science and scientists. In fact, Bacon's *New Atlantis* is rife with religious images of progress brought about by the new priests of science and heavily influenced by the millennial thought of Joachim of Fiore (among others).[32]

There are several ways in which science adopts, within its very structure, religious ideas. First, it adopts the human exceptionalism (see chapter 7) that was read into the Christianized Ptolemaic universe. In other words, just as human beings were created in the image of God and as the center of salvation history over and against the rest of the natural world, now humans are the locus of value and the rest of the world is instrumental to scientific goals of progress. Second, the very idea of revelation and a good creation that was ordered by God is translated into universal natural laws that can be discovered through the light of reason. In other words, the book of scripture takes a back seat to the book of nature during the emergence of modern science. However this book still relies on a nature that is predictable, stable, and not out to trick us. Modern science could have never emerged from the radical skepticism that Descartes's posits with his Evil Genius experiment. It must assume that the world operates according to laws, and such laws were, at least at first, assumed to be set in place by an original Creator. There are not dueling gods of the material and spiritual worlds, but a single creator God that lays down laws of nature and gives human beings reason to understand those laws. This leads me to my third and related monotheistic assumption of modern science: monotheism itself suggests that there is one, universal truth. This understanding of truth becomes the justification for scientific laws and theories. In other words, unlike religion, science becomes acultural and objective: it is thought to be the same everywhere. This form of universalization is every bit as colonizing as religious forms of universalizing. Science takes over the universalizing tendency directly from the universalizing tendencies of a monotheistic God that creates everything out of nothing. It is this universalizing process that shapes the contemporary model of globalization and it always already includes both science and religion.

My point in arguing this is that science and religion have always been and are necessarily in dialogue. To deny this is to turn religion into a personal, apolitical matter and to turn science into a logical, universal, apolitical matter. Depoliticizing religion and science is a myth: they are always already political. This is just as true today as it was during the shifting of worldviews five hundred years ago. An apolitical religion and science are precisely what allows for the solipsism that happens in foundationalism and circular thinking that I argue are dead-end options for justifying knowledge claims. In other words, apolitical science or apolitical religion provide "that which you cannot get behind" arguments for justifying knowledge. Such arguments, whether circular or foundational, begin to create inertia in thinking and practices, and such inertia eventually begins to order the entire planet in its own image. Through denying the political nature of science and religion, we ignore the ways in which modern Western, secular culture is inherently shaped by a specific religion and thus, in globalizing modern science and culture, we colonize the rest of the peoples and places in the world. As the reader is well aware, globalization is not about a process of equal exchange between cultures and places; it is more aptly described as "globalatinization" or the imposition of Western ways of knowing and being on the rest of the entire planet.[33]

The Myth of the Secular and the Space of Capitalism: Globalatinization

As a part of the process by which religion becomes personal and private and science becomes objective and public, science and economics become handmaidens in defining public spaces. In other words, science and economics (of the free-market-capital variety) eventually become that which is relegated to the realm of secular space, along with legal systems. Law, science, and economics are then supposed to transcend personal and individual religious preferences. However, this very secular positioning of the places of science and religion itself harbors religious underpinnings. First and most obvious are the very separating out of these realms from overall culture and life. Something that only happens as a result of the development of the concept of "secular" in Western, Christian thought. The etymology of the term is found in the Latin *saeculum*, which was used to designate a period of roughly one hundred years and in distinction from the eternal time of the sacred. "*Secular* referred to the affairs of a worldly existence and was used in the middle ages specifically to distinguish members of the clergy, who were attached to religious orders from those who served worldly, local parishes."[34] In other words, it was a term that was meant to distinguish between worldly and spiritual things, something that is especially unique to the context of monotheistic cultures. It in no way had the connotation that it does today between the radical separation of secular (public, reason, state) and sacred (personal, faith, church). Such a process of "disenchantment" and its accompanying immanent understanding of the world is what Charles Taylor addresses in his massive undertaking *A Secular Age*. In this book Taylor traces the secular as a state in which "Belief in God is no longer axiomatic. There are alternatives. And this will also likely mean that at least in certain milieu, it may be hard to sustain one's faith. There will be people who feel bound to give it up, even though they mourn its loss."[35] Though this is not the place to argue with Taylor's massive tome nor dive into it, I must admit to certain points of agreement and disagreement. First of all, I agree with him in terms of the process by which the secular becomes natural through the mechanization of the rest of the natural world and the rise of modern science. I would, however, not refer to this as a disenchantment, but rather a different enchantment: one that sees nature as a mechanical order devoid of the divine. Second, I agree with him that "secularism as a philosophy of history, and thus an ideology, is to turn the particular Western Christian historical process of secularization into a universal process of human development from belief to unbelief, from primitive irrational or metaphysical religion to modern rational post-metaphysical secular consciousness."[36] In this way the myth of the secular is a universalizing myth. It is being imposed upon multiple other places, peoples, and spaces across the globe in a monological manner. However, I would not characterize this myth as solely the result of an "immanent" frame. Such a juxtaposition between the immanent and transcendent frames is yet another contextually Western juxtaposition. In other words, as William Connolly suggests, Taylor's reading of "immanence" as somehow closed is due to his theistic understanding of reality, which posits transcendence as the open space of immanent reality.[37] In fact, it is possible to imagine an open and becoming immanence. From this open immanence, it is in the very collapse into immanence from which so many postmodern thinkers argue, that some form of spirituality or religion is allowed to return. As Bruno Latour notes, "Deprived of the help of transcendence, we at first believe we are going to suffocate for want of oxygen; then we notice that we are breathing more freely than before: transcendences abound" in the possibilities that emerge when we get away from categories that rely on acontextual transcendence.[38] In other words, and as I argue throughout the book, the foundational categories of reductive materialism (in science and nature) and robust theism or idealistic certainty (in religious thought and meaning-making practices) are actually a collapse into a flat world where everything is ordered. It is the doing away with these transcendent foundations that allows for immanence to become an evolving space of multiple possibilities. In addition to this point, other cultures in which immanent spaces and some amount of pluralism have existed do not exhibit the type of secular/sacred (nor immanent/transcendent) split that Taylor describes.

Many religious traditions and the cultures they are a part of do not separate the economic, legal, and scientific from the religious. Let me offer a brief example here, that of Jainism. In Jaina philosophy one finds cosmological, biological, chemical, and neurological among other descriptions of the world. These are not understood either as in dissonance with the realm of public Western modern science or as somehow private belief. Rather, they are taken together with modern science as various ways of understanding the world around us. The Jain principles of *anekanta* (nonabsolutism) and *syadvada* (relativity) are also relevant here. These two doctrines suggest that there is no one right way of understanding the world around us and thus no one correct description of reality. Rather, there are multiperspectival approaches, and these multiple perspectives do not include only human ones. Many perspectives give us different insights into a larger reality around us, yet none can exhaust the full meaning or truth of the realities we inhabit. Such a position is not only fruitful for multifaith dialogue but also challenges the idea that somehow religion and science are on different footing. It is not just that religion and science offer different perspectives on the reality around us, but rather that they both offer descriptions of the natures that we emerge from and inhabit. It is not just the old constructivist idea that there is one nature and many cultures, but rather that the world is inherently multiperspectival.

We can talk about multiple natures: a Jain understanding of nature, a Western scientific understanding of nature, a Buddhist understanding of nature, etc. These different perspectives then come with many different possibilities for acting in the world according to those perspectives. In other words, they come with regimes of technology that help shape the worlds according to given understandings. As Talal Asad notes, modernity too "is a *project*—or rather, a series of interlinked projects—that certain people in power seek to achieve."[39] As such, we ought to begin to understand that which we call the "secular" or "modernity" as but one among many competing truth regimes rather than as some inevitable force of history. In fact, the secular, scientific world of modernity itself has a primary paradox this book will challenge, that of the liberal individual. "The paradox inadequately appreciated here is that the self to be liberated from external control [i.e., to be liberated from the darkness of belief and superstition through enlightened reason, modern science, and secular governments] must be subjected to the control of a liberating self already and always free, aware, and in control of its own desires."[40] Such a self, and the categories of transcendence that self depends upon (self/other, human/nature, agent/patient), only exists when one backgrounds the many different others that make up the self (see chapters 4–5 especially). A better option, or so I will argue, is that we understand the secular worldview as but one of many truth claims. It probably has something to offer, but forcing Western-style secularism upon the face of the globe amounts to colonizing others and a religious war on other's beliefs.

Recognizing that one's perspective is both unique and ecohistorically located does not mean that it is relative and meaningless, but rather that it contributes a voice to the overall becoming process of an ongoing multiperspectival planet. One can be happy to contribute one's voice to the mix without suggesting that it is the voice of universal law. One can argue persuasively for one's own point of view without being dogmatic about holding onto that point of view. One can see many different possibilities for planetary becomings, choose some of those possibilities to act upon, and then take responsibility for the consequences of those actions. From such a perspective, there is no pure science nor pure religion, but rather there are multiple persuasions / (per)versions of religions, sciences, and their understandings of nature.[41]

From this planetary epistemology, science and religion are not so much related through dialogue, nor are they in conflict or consonance, but rather they are multiply informed, located perspectives on the world around them. To start from a pure secular science and approach a pure personal religion is itself a mistaken way of relating the two (whether arguing they are in dialogue, consonant, in conflict, or that they describe "two worlds"). The point here is that they always already inform *any one* of the multiple perspectives on reality taken. It is hard to understand the worlds we inhabit without the "taking apart" of the scientific method and the "gathering" of meaning-making practices. Thus, if one mistakes her perspective with reality, it is a form of what Jainism might call conceptual violence. This conceptual violence seeks to eradicate the *différance* of deconstructionism through conceptual conformity.[42] Such forcing of conceptual conformity is only possible, I argue, through the solipsisms of foundational and circular thinking. The globalization of modern Western science and its handmaidens in the secular space (free-market capitalism and industrial-style development) is one such solipsism that is being argued as the only way forward for planetary becoming. It is seen as the only viable option forward for the processes of globalization, yet it is really the imposition of one perspective over the face of the entire planet. In other words, the command to open up one's country to the globalization of free-market capitalism is very much a religious and personal issue. Second, and related, just as modern science harbors its own religious underpinnings of belief in progress, understanding nature as dead matter that can be transformed, and anthropocentrism, so it is with modern understandings of economics.

I need not rehearse here the Weber thesis and other connections between capitalism and the Protestant religions, or the role of the Reformation in developing a work force for the industrial revolution, but rather, I will focus on the theology behind John Locke's understanding of property, which is still the definition of property that justifies free-market capitalism.[43] For most educated in the history of modern Western economics, its secular description begins with the *Second Treatise* of John Locke and the definition of private property he lays out there. That is, private property is the result of the individual human mixing his labor with dead matter. Later, and of course through industrialization, the wealthy are also able to gain more capital through the use of others' labor—something Locke's own theory did not support. It is this very notion of an individual mixing his own labor with dead matter that is at heart religious.

What most people don't read about in this secular history is that Locke's *First Treatise* provides the theological underpinnings for such an understanding of the individual human as active and nature as dead matter.[44] Locke was writing in the time of the Glorious Revolution and against the work of Robert Filmer. Filmer interpreted the language of Genesis regarding the *imago dei* in an aristocratic way suggesting that some humans could have dominion and rule over others.[45] Locke, on the other hand, argued that all humans shared dominion equally over the rest of the natural world. He democratized the *imago dei* and the dominion clause of Genesis in such a way that made everyone their own little rulers. Further, the cosmology that provided his understanding of nature as dead matter was being developed by Descartes and Newton. Thus some form of monotheistic deism underwrites the anthropology and understanding of nature that leads to the development of Locke's private property, which then with modification becomes the basis for free-market capitalism (not to mention individual human rights and liberal understandings of freedom). Capitalism and science, then, share the understanding of nature as dead and humans as over and against that nature. And, I argue, these are religiously loaded rather than secular concepts.

If we skip over the first wave of colonization and enter the era of the contemporary globalization of (a heavily subsidized) free-market capitalism, then we can begin to see how these notions of the secular can be interpreted as religious attacks on other religious traditions and peoples that do not understand nature as dead, nor humans as little gods capable of creating their own worlds out of nothing (or dead matter). We must then trace the construction of secularism through a project that includes an *ex nihilo* creator God, Renaissance humanism, the Enlightenment concept of nature as dead matter, the idea of history as single (Hegel), and in general to some very Christian and Western ways of understanding the world.[46] It is a further offense to "others" when the religiosity of these concepts is covered over as secular or as business as usual. There is a sense then in which the spread of capitalism and modern science is, indeed, a religious war of planetary proportions. But within the very walls of modern science itself, and perhaps as a result of the mixing of identities through the process of colonization and globalization, lie some keys toward breaking down the logic of domination found in globalization. New postmodern sciences are beginning to challenge Christianized understandings of anthropology and nature as dead matter. Excluded others in the process of globalization have now returned to redefine and reshape the center's self-understanding. In the next two chapters I will describe this process under the headings of "Destabilizing Nature" (chapter 2) and "Destabilizing Religion" (chapter 3).

2 DESTABILIZING NATURE

Natura Naturans, Emergence, and Evolution's Rainbow

> Both scientific realism and religious fundamentalism are private projects which have got out of hand. They are attempts to make one's own private way of giving meaning to one's own life . . . obligatory for the general public.
> RICHARD RORTY, *Philosophy and Social Hope*

As mentioned in the previous chapter, the always already of religion and science together is covered over in attempts to make one side of the religion and science equation the foundation by which the other can be explained. Most often this is accomplished through stories about what nature or ultimate reality is. Such stories are created through histories of cultures and their institutions reflecting upon the type of knowledge that is under investigation and the methods for gaining that knowledge. These practices all too often become foundational and cover over the process by which they were constructed over time. When this happens, the stories we tell ourselves serve to reify all of life into a particular natural-cultural narrative. Both scientific discourse about nature and religious discourse about meaning and value are guilty of this process of reification. Science aids this process of reification or "misplaced concreteness" because it becomes a transcendent source of what is natural.[1] Furthermore, because the natural sciences, according to some materialist and reductionist accounts, grasp and manipulate a knowable, static, and stable nature, formulated by natural laws, they become the source of hope for a better human future. If we can just get better knowledge, closer representation of our theories and equations to reality, then we will be able to control that reality. Hope for a sustainable future, for instance, relies heavily on developments in science and technology. In this way, as Žižek notes, science now plays the role of providing certainty where religion once did. He writes, "The paradox effectively is that, today, science provides security which was once guaranteed by religion, and, in a curious inversion, religion is one of the possible places from which one can develop critical doubts about contemporary society."[2] Perhaps Žižek gives too much credit to religion here, but more will be said about the religion side of the equation in the next chapter. For now let it suffice to note that religion too can aid in the process of projecting linearity onto the rest of the natural world (and even the whole universe). Stories of creation that provide universal understandings of meaning in the form of movement from *arche* to *telos*, take human beings out of the depths of their interactions with the rest of the natural world and background that which does not fit in with the narrative. This is true whether we are talking about the soteriological narrative of Christian theology or Teilhard de Chardin's narrative of a universe evolving toward an Omega Point[3]—or even the *Universe Story* moving from the big bang toward ever more complex forms of life.[4] The examples of the Dark Ages written over by the light of the Renaissance and Columbus's accounts of his discovery of the "other" world serve well to illustrate the ways in which both science and religion partake in this linearizing of life on the planet and reveal the violence that such linearity creates. As with all systems of power, these stories are successful because they hide their own mechanisms of production and maintenance.[5] That is, they hide over the ways in which their norms, truths, and knowledge are constructed through mechanisms over time to seem, well, natural.

In destabilizing nature, this chapter argues for some alternative, disruptive stories that all rely on some sort of notion of radical immanence.[6] Radical immanence in the realm of nature will mean some version of what Spinoza calls *natura naturans*, or nature naturing, without *naturata*, or nature natured. That is, a nature becoming in many different possible directions but in no preordained direction and with no essential natures. As Timothy Morton suggests, becoming nature as the totality of our reality (including thoughts and meaning-making practices) "doesn't mean something closed, single, and independent, nor does it mean something predetermined and fixed; it has no goal."[7] There is no "natured," but only the continual process of nature becoming. This understanding of an immanent and ongoing nature provides a *viable* option for redefining nature as a transformative political space-time of planetary possibilities rather than as transcendent source for foundational claims. Karen Barad also has such an understanding of nature from the perspective of Niels Bohr's analysis of the quantum world. More will be said on this later in chapter 2.

Though I am separating out the destabilizing process into two chapters here (destabilizing nature and destabilizing religion), religion and science, values and facts, meaning-making practices and nature are always already together. In fact, it is in part the meaning-making practices inherent in sciences and science's effects on meaning-making practices that destabilize these categories from becoming foundational. This chapter will, then, not only destabilize the science of nature, but destabilize the category of nature by uncovering the inherent meaning-making practices in its construction.

Particularly what we understand as the sciences of evolution, ecology, Einsteinien and post-Einsteinien physics, chaos and complexity, and quantum physics and cosmology are all influenced by an influx of Hindu and Buddhist thought into the Euro-Western world during the seventeenth through nineteenth centuries. "Except for ceramics, philosophy was in the seventeenth century probably the most important cultural 'import' from the Far East to the West."[8] Furthermore, many twentieth- and twenty-first-century practitioners of these sciences deeply and overtly examine Taoism (e.g., Fritjof Capra), Confucianism (e.g., Terrence Deacon), and Buddhist and Hindu ways of meaning-making practices. Precisely because these thought systems focus on interconnection and process rather than essence and substance, they have contributed to new regimes of truth for understanding the natural world.

Lest one accuse me of an idealistic orientalism here, we must also acknowledge the counternarratives to substantial and transcendent thinking even within Western histories. There have been many mixings of thought in the mythical West that draw from more immanent and interconnected understandings of reality, not to mention the fact that ancient cultures and religions are always already influenced by contact with others through trade routes and the sharing of knowledge. Hence it is difficult in the end to even try separating thought into categories of East and West or indigenous and modern. Rather than understanding these differences categorically, here I understand them as more like trends or streams of thought, some of which are more substance-based and others of which are more processive, regardless of the origin of the idea. My main point here is to show that ideas shape matter and cocreate worlds just as much as matter shapes ideas and cocreates worlds. Hegelian idealism and Marxist materialism are but two sides of the same coin. A stress on a radical immanence that keeps ideas-matter on the same plane of existence keeps us focused on the processive, evolving process of identities formed in relationships, and in and through difference (which is the subject of chapter 4). Such identity formation occurs not just in regards to human identities, but rather the "performativity" of identities goes "all the way down." As Barad notes, "Matter is produced and productive, generated and generative. Matter is agentive, not a fixed essence or property of things. Mattering is differentiating, and which differences come to matter, matter in the iterative production of different differences."[9]

Radical immanence, as you will see in my discussion of Bruno, Spinoza, and the emergentists means that religion is also a part of nature as a natural-cultural human projection. Humans, in this understanding, are a meaning-making part of the rest of the natural world. As such, our meaning-making capacities emerge from natural-cultural contexts and return to affect the rest of the natural world. Religion as a meaning-making practice provides a space in which to imagine possibilities for a viable planetary community. In other words, religion as a located meaning-making practice, *as precisely of and for the natural-cultural worlds we inhabit*, may provide a transformative political space for the type of planetary existence Žižek wants, rather than serve as the transcendent foundation for universalizing meaning. In fact, I will argue that when we understand religion and science together, we will see that changes in meaning-making lead to changes in nature and vice versa. In this chapter I will look at this in terms of nonequilibrium thermodynamics and other shifts in physics that uncover a reality that challenges the dominant substantial metaphysics in Western meaning-making practices and the reductive materialism of modern science. As such, modern (mechanistic) science is reconceived as yet another traditional ecological knowledge arising from a specific place, with specific values, and cocreating specific worlds. Indeterminacy means that truth is not representational, but functions more like performative truth regimes. Whatever science calls "natural" reveals itself in part as a result of the apparatus investigating the specific phenomena and the natural-cultural location of the scientist. "Apparatuses are not mere observing instruments but boundary-drawing practices—specific material re(configurings) of the world—which come to matter."[10] In other words, the observing apparatus is part of what is being observed just as our subjectivity is part of the observing apparatus.[11] This does not mean that we humans create our own realities, but rather that we are created by and return to cocreate realities based upon the emergent meaning-making processes that make up a part of our own emerging, always becoming identities.

This realization means that we are entering a period in which there is a new death of nature, separate from that described by feminist environmental historian Carolyn Merchant and that described by environmental writer, Bill McKibben. For Merchant, the death of nature meant the process by which modern science makes the world "dead matter."[12] Though this is one way of understanding the world, I would argue that this is no less enchanted.[13] Indeed, the enchantment of making the world dead matter is found in the marvels of modern technologies that such a mechanistic truth regime ushers in: the wonder and marvels of skyscrapers, space travel, air travel, the Internet, and the very sciences that emerge out of the mechanistic model of science (even if those sciences contain the ultimate demise of mechanism) are all quite enchanting.

For McKibben, the death of nature means the death of the idea of a nature that is untouched by human hands.[14] The idea of such a nature, or its possibility, belies belief in at least a past separation between human culture and the rest of the natural world. One could argue that there was such a nature before the emergence of *Homo sapiens sapiens*. But, this is as fruitful as imagining a nature before the emergence of alligators, dolphins, trees, or any other life on the planet. In other words, of course, humans have not always been a part of nature, but to say that once humans emerge *from* nature that they can kill nature as a whole simply gives humans too much power and leads to the question of what humans are if they are not part of the process of nature naturing.

The death of nature I articulate here is the death of the very idea that there is something like nature that humans, cultures, and religions are somehow not a part of. Such notions have become destructive for planetary becoming because they perpetuate the myth of essential identities in identity politics (chapters 4 and 5), environmental ethics based upon place (chapter 6), and human exceptionalism (chapter 7). This version of nature with its identity politics, ethics of place, and human exceptionalism is nothing but "the capitalistic fantasy" or nature as something to be consumed.[15] This new death of nature is actually a *queering* of the boundaries between humans-technology-animals and the rest of the natural world in the past, present, and future. For this reason, I end this chapter with a discussion of the implications of queer theory for understandings of nature. Queer theory, however, does not develop ex nihilo and, I would argue, has conceptual kin in concepts of religion and nature that rely on some sort of radical immanence. It is to these historical trajectories that I now turn, first in the works of Bruno, Spinoza, and the early emergentists, then into what have been termed by some as postmodern sciences.[16]

Bruno, Spinoza, and the Emergentists: Stories of Radical Immanence

> The philosophy of nature itself, in other words, is no longer grounded in somatism, but in the dynamics from which all ground, and all bodies issue: "matter is precisely just matter, that is, the ground of bodies but immediately therefore, not corporeal."
> IAN HAMILTON GRANT, *Philosophies of Nature After Schelling*

Radical immanence has been a tradition or trajectory over and against metaphysics of transcendence and substance for quite some time. Instead of the linear narrative of modern understandings of nature that begins with Copernicus and the scientific revolution, and still depends heavily on the metaphysics of transcendence and substance, what if we highlighted the story of Bruno's immanent and infinite universe? Though not suggesting a linear origin for the present way of thinking about nature, beginning with Bruno provides us with the space where multiple universes can be theorized: it upsets the linear narrative that puts Copernicus and Galileo at the origin of the modern scientific world. Just as queer theorists depend on finding queer identities in the past that they then deconstruct, and liberation, feminist, and race critical theorists of all sorts depend on identifying the hidden voices of their subjects throughout history—even if denying the idea that there is an essence to any of those subjectivies—so I attempt here to bring together some of the voices on immanence throughout Western history. Such stories and identities together begin to show us that we are and have always been much more diverse than dominant hegemonic narratives of Western thought suggest.

In the brief sketch here I will follow a line of developing a tradition of radical immanence from Giordano Bruno to Spinoza and then suggest that these were the grounds from which many contemporary rereadings of religion and nature began: from American pragmatists such as James and Dewey to French constructive postmodernism, such as is found in Deleuze and Guattari, and the contemporary discussion of emergence theory. Such rereadings as found in these immanent systems of thought do not emerge ex nihilo. Rather they are part of natural-cultural traditions that can be found in "the Far East" and from nonsubstantial thinkers in the ancient Greek and proto-Western worlds.

The Multiverse of Bruno

> Dissolve the notion that our earth is unique and central to the whole. Remove the ignoble belief in that fifth essence. Give to us the knowledge that the composition of our own star and world is even as that of as many other stars and worlds as we can see. Each of the infinity of great and vast worlds, each of the infinity of lesser worlds, is equally sustained and nourished afresh through the succession of his ordered phases. Rid us of those external motive forces together with the limiting bounds of heaven. Open wide to us the gate through which we may perceive the likeness [52] of our own and of all other stars.
> GIORDANO BRUNO, *On the Infinite Universe and Worlds*, fifth dialogue

Giordano Bruno (1548–1600) has been often used as the example par excellance of the rift between science and religion during the Western scientific revolution. Indeed, this is a position that both obscures the historical realities (Bruno was, after all, a Dominican monk his whole life and an avid reader of, among others, Aquinas, Aristotle, and others "accepted" by the church) and obscures the emergence of another possible interpretation of Bruno for narrating multiplicity in the face of the increasing imposition of sameness over the globe. "Bruno held with Nicolas of Cusa that an infinite universe can provide no absolute position, no centre or circumference, but that the position of our world or of any one object within or without it can be defined only in relation to another object."[17] Such an insight would not be taken up scientifically until Einstein did so some three hundred years later! Bruno's context was one of expanding information from multiple sciences, discovery of new worlds and the Renaissance of the ancient Greek world, undoubtedly influenced by what we might call Eastern thought. Note I am not arguing for a romanticized orientalism, but rather for the idea that figures such as Nicolas of Cusa and Bruno, through their reengagement with some of the more esoteric ideas in Greek thought, paved the way for thinking about the universe differently in ways that were always already influenced by Eastern ideas.[18] Rather than opting for the imperial, universal interpretation of reality as usual, Bruno's rather nonsubstantial and non-Western ways of thinking led him to opt for a multiverse. Bruno's "philosophical system privileges difference and heterogeneity over unity and harmony. The model of empire needs to homogenize. Bruno loves the plural moment."[19]

For Bruno, of course, nature includes both material and soul. That is, God is nature in that the world is the infinite, becoming process of God. All that is in nature, including the universe, is God becoming.[20] Bruno also considers the universe to be infinite in its becoming, rather than merely heliocentric. He not only displaces the earth, but the sun as well as any other center. This understanding of nature suggests a community of biohistorical creatures becoming toward many different possible directions. Furthermore, it suggests a universe that is moving toward infinite possible manifestations: a plurality, a multiverse, or a multinaturalism. Thus perhaps in Bruno it is also possible to read humans as meaning-making creatures, nature as nature naturing, and the future as radically open.

Is it possible that Bruno's development of a pantheistic multiverse might be the result of the always already of Eastern and Western thought that was "rediscovered" with a passion by the Renaissance monks? Is it also possible that his is a model for understanding the always already of religion and science inherent to our meaning-making practices rather than as the example of religion versus science?[21] I think the answer to these questions is a resounding yes: figures such as Bruno and Cusa paved the way for the reflux of Eastern thought into Europe (largely through the Jesuits) and to the eventual influx of thinking that helped lead to nonmechanistic understandings of the rest of the natural world. In any event, one such thinker, deeply indebted to the thought of Bruno, who developed a full-blown pantheistic worldview that has been compared to the Vedanta traditions of India and was a major source for the Romantics, who drank deeply from the wells of Eastern thought, was Baruch Spinoza.

Spinoza's Pantheism: Natura Naturans (Who Needs Naturata?)

Bruno's work had a huge influence on the pantheistic understanding of the Amsterdamer Baruch Spinoza in the seventeenth century.[22] For Spinoza, "God must be immanent in the natural order, the creator in its creation, if we are to avoid the incoherence of thinking of two substances in reality: a creator distinct from his creation."[23] Nature or God (according to Spinoza) is of one substance and is in constant flux.[24] "By *Natura Naturans* we must understand what is in itself and is conceived through itself, or such attributes of substance as express an eternal and infinite essence, that is, God."[25] Nature naturing is the sense of active God/Nature becoming. It is true that Spinoza speaks of God/Nature as a single substance. A diversity of becoming is simply not in his seventeenth-century vocabulary. However this radical immanence does beg the question of why? Why and how does he also hold on to a natura naturata, nature natured, if there is one, becoming substance? Doesn't such a nature natured imply some sort of transcendence from the one natura naturans?

Spinoza speaks of nature natured as that which has been created in the past. Of course, what follows by way of an answer to why he holds on to this transcendent notion is speculative, but it seems that we could with good reason think of Spinoza as bridging the idea between a world dominated by substance thinking and his pantheist, immanent world of nature naturing: just as Cusa and Bruno before him bridged the worlds of a single historical reality with multiple worlds and realities. In fact, many argue that Spinoza's pantheism, indebted to Bruno and Cusa, was also developed in relationship to his readings of Confucius, who was translated into Latin by that time.[26] Thus his philosophy of immanence was also much more attuned to naturalistic explanations of reality: again, not necessarily the reductive science of Newtonian mechanics, which requires an objective transcendent source, but a science of understanding reality immanently. "Spinoza was" among "the first to accept the results of natural sciences of our modern times and to build upon these fundaments the structure of a new faith which Santayana once quite appropriately called a religion of science."[27]

Spinoza's concept of nature natured seems to be in line with a more reductive mode of science that understands matter as dead substance rather than an immanent understanding that brings spirited matter together in an immanent understanding. Perhaps Spinoza's nature natured is more about persuading and luring people into his way of thinking rather than an *essential* component of his otherwise immanent philosophy. Speaking from and during a time when mechanism was the ruling metaphor of nature in the European West, Spinoza did well to equate God with that understanding of nature: even if it meant maintaining some sort of transcendence in the concept of nature natured. Whereas Bruno's radical notion of becoming fell on deaf ears and landed him into some trouble, Spinoza's bridging of substance and becoming became and still is quite influential. He was in his context and rather than being a lonely voice. The concept of nature natured provided a persuasive bridge toward thinking radical immanence.

Thus here I want to entertain the idea that natura naturans is all that there is: nothing is ever natured in the sense that it has a created, identifiable essence.[28] The very idea of nature natured may then be understood as a persuasive hangover from substantial ways of thinking. According to contemporary sciences of quantum reality and nonequilibrium thermodynamics, there truly never is a part of life that is not continuously interacting with other parts: even in death. All is becoming other with the others of nature naturing. All identities, whether human, plant, animal, mineral, cellular, quantum, are porous and in constant interaction and exchange. Without this exchange and interaction life would not move from moment to moment and the universe would be frozen in time.

One can begin to hear traces of the assemblages and rhizomes found in the work of Deleuze and Guattari.[29] For Deleuze and Guattari, as for Spinoza, nature is nature naturing. Arboreal or foundational understandings and metaphors for thinking about identities really don't capture the movement and flux of nature naturing, so they posit the rhizome as a nature-based, radically immanent metaphor for thought. A rhizome has no beginning or end and can shoot off in various directions. So it is with nature: humans and our cultures and technologies are but various "lines of flight" from within the rhizomatic cosmic expansion and planetary evolution. Thought and matter can, much like a rhizome, move in multiple directions without an inherent point of origin or final goal or end.

Again for Spinoza nature is open and humans are meaning-making creatures. He writes, for example, "Not many words will be required now to show that Nature has no end set before it, and that all final causes are nothing but human fictions."[30] In his understanding of natura naturans there is caution against forming a holistic foundation as well. "Because human beings can learn to appreciate the world as a single, divine whole of greater worth than any individual element within it, they themselves," in this line of thinking, become "judged of greater worth than any other individuals."[31] Thus it may be that we can speak of an open, evolving nature in Spinoza's pantheism, rather than some sort of wholism for which some deep ecologists have argued. The difference is that in an open, evolving nature "we can never have knowledge of nature as a whole. *Divine* power is manifest in its diverse parts (places, species, lands, and waters); each *self* is not identical to any other; and every body comprising the *world* consists of diverse matter."[32] As Deleuze notes in his own study of Spinoza, this freedom is the radical freedom that only a collapse into immanence can supply: one is finally free from the external impositions of ultimate origins, ultimate ends, and other modes of foundational thinking. "When Spinoza says that we do not even know what a body can do, this is practically a war cry. He adds that we speak of consciousness, mind, soul, of the power of the soul over the body; we chatter away about these things, but do not even know what bodies can do. Moral chattering replaces true philosophy."[33] Such an embodied philosophy was influential on critical theory (much later), but also on the early emergentists of the nineteenth and twentieth centuries and even the quantum thinkers of this period as well. One early taproot for thinking about emergence was Henri Bergson.

Henri Bergson and the Élan Vital

Yet another source in the nineteenth and twentieth centuries for a destabilizing understanding of nature is found in Henri Bergson and the early emergentists. The early emergentists, similar to the pragmatists, were trying to think of how ideas, culture, and mind evolved from the rest of the natural world. As such, they began to think of our existence and our knowledge of that existence as without ultimate origins or final ends and in ways that were nondualistic (again, probably influenced by the nondualism of nonsubstantial metaphysics found in much Eastern thought and influenced by thinkers such as Heraclitus and Spinoza in the West—though these thinkers both already represent hybrids of Eastern-Western thought). Bergson argues, "We must get beyond both points of view, both mechanism and finalism being, at bottom, only standpoints to which the human mind has been led by considering the work of man."[34] In other words, Bergson, like Spinoza before him, is trying to make sense of a world that has collapsed into immanence. In particular, he is trying to make sense of evolution in a way that is not reductive, but allows for something like the emergence of mind to bubble forth. He finds a way of doing this through opting to look at life nonsubstantially and phenomenally. "There are no things, there are only actions."[35] For Bergson, as for emergentists today, the story of evolution is the move from "something more from nothing but." In other words, emergentists are curious about how new things come into existence from all that is. Bergson posits the "life force" or *élan vital* as that creative striving in life that leads to the emergence of new things. Such an explanation is surely shaped by his pantheism and more so by his nonsubstantive ways of thinking, moving between the dualism of materialism and idealism. Perhaps I overstate the flow from the always entangled East-West thought among the ancient Greeks to Cusa, to Bruno, to Spinoza, to Bergson, but we do know that Bergson was well-read in Eastern texts. He mentions Eastern thought in his works on religion, especially the mystical components of religious traditions.[36] My point is that these ideas that life is primarily about action, exchange, and in general nonsubstantial, begin to reflect the ways in which thinkers see the rest of the natural world: there is a move from thinking in terms of the truth regimes of reductive science toward that of what we might call a nonreductive emergentist science. Before moving along to the science of Einsteinian and post-Einstinian physics, I first want to discuss another source for nonreductive thinking about a multiple-emergent planetary understanding of reality: the pragmatists. Not only did William James and Bergson work together, but the pragmatists in general drew from Bergson and many of his predecessors in articulating a "radical materialism."

Radical Materialism, Pragmatism, and Habits of Nature-Culture

As mentioned earlier, American pragmatists, French postmodern philosophies, and the contemporary discussion of emergence are much indebted to this tradition of radical immanence.[37] Pragmatists, especially, have suggestions for the role of humans within nature naturing as meaning makers. Through the emergence of habits, humans make sense of the worlds around them. "Habits constitute morphogenetic fields that make it easier for subsequent generations to accomplish something that was more difficult for earlier generations."[38] However these habits of meaning-making processes do not mean that any one way of being is right, nor that any one knowledge of reality represents reality any more precisely. "Biological evolution produces ever new species, and cultural evolution produces ever new audiences, but there is no such thing as the species which evolution has in view, nor any such thing as the 'aim of inquiry.'"[39] Such thinking flies in the face of human exceptionalism and any notion of the anthropic principle, which will be discussed further in chapter 7.

Pragmatic truth is similar to Bruno Latour's understanding of truth where neither scientific nor religious "truth is . . . to be found in correspondence—either between the word and the world in the case of science, or between the original and the copy in the case of religion—but in taking up again the task of *continuing* the flow, of elongating the cascade of mediations one step further."[40] Again, humans are meaning-making creatures, the future is open, and nature and religion are wed through a metaphysic of radical immanence.

In taking this language of habits one step further, perhaps we can also begin to think of nature itself as creating various habits. On the biological level we can speak of evolutionary habits of becoming, such as genetic mutations, morphological developments that shape life toward becoming in specific ways, and cellular tendencies that lead to certain functions. We can also speak of habits on the chemical level: the probability of chemicals reacting in a given way depending on their elemental structure and makeup; the probability of a molecule connecting with another to form a new quality, such as the quality of H_2O that becomes buoyant only when connected with many other H_2O molecules. The physical and cosmological level can also be spoken of in terms of habits: the rate of cosmic expansion, gravity, nonlocality, and thermodynamics all display probabilities. The benefit of thinking of nature naturing in this habitual way is that it recognizes that these habits can change, they can lead toward new ways of becoming: even gravity breaks down at the edge of the cosmos, which is why the cosmos is expanding rather than contracting. Ecosystems, as habits of becoming, evolve and change through their ecotonal edges and shifts in climate. Such a suggestion destabilizes the concept of nature in that it does not become the source for the way things are, but rather the way things have been and tend toward. There is nothing unnatural in suggesting that these laws could morph into something different or even that sometimes chemicals, genes, and molecules don't behave in predictable ways. From a perspective of radical immanence, one in which we don't stand outside of nature looking back at the whole 13.7 billion year cosmic expansion, tendencies and habits are a much more accurate way to describe nature than are universal laws. Barad agrees with this sentiment and extends the notion of performativity to the quantum level in her discussion of "agential realism." She writes, "according to agential realism, causality is neither a matter of strict determinism nor one of free will."[41] Holding on to determinism or free will would mean that either humans are collapsed into a deterministic, reductive mechanism or that the rest of the natural world would be treated anthropomorphically as having individualistic free will. Rather, her notion of performativity, much like *habitus*, extends "all the way down" (and "all the way up" for that matter). If we begin to understand natural laws, and cultural norms as habits rather than as somehow natural or something that dictates life from above, we can also begin to try to break bad habits and work toward truth regimes that support habits of planetary flourishing. We can promote alternative performances in our coconstructions of the world. Not in a way that makes all of nature a "standing reserve" for human beings, but in a planetary, multiperspectival, nonanthropocentric way (see chapter 7).

There is no more simple equation between natural = good or natural = bad, for that matter. Cancer and AIDS are just as natural as love and happiness, the forest is just as natural as the shopping mall, and solar energy is just as natural as nuclear energy. Assuming that what is human made is unnatural not only participates in human exceptionalism, but it suggests a transcendence of the present where past is natural and future is engineered or technological. Such thinking relies on a present transcendence and an assumption that human thought is somehow not part of the emergent and evolutionary process of nature naturing.

From this immanent perspective we begin to identify those habits we understand as bad, and work toward reform: a cure for cancer and AIDS, prison reform away from incarcerating peoples based upon racial profiling and years of racial injustice, universal health care reform, and creating energies that don't leave future generations of life on the planet with radioactive toxic wastes to fear. Much as we try to cultivate good habits of exercise, so science and technology ought to be geared toward promoting good habits for the future of planetary becoming (the subject of chapters 5–7).

Contemporary theories of emergence also draw from this nonsubstantive, immanent thought in at least three ways: there is one substance (a type of radical imminence); humans, as Deacon notes are *The Symbolic Species* (or meaning-making creatures); and the future is open (autopoesis, emergent newness). Each component deserves more elaboration.

The contemporary discussion of emergence draws from the early emergentists such as Henri Bergson, but tends to be more focused on how "something more" emerges from "nothing but" rather than the somewhat idealistic notion of a life force or *élan vital*. In other words, contemporary emergentists are much more about a radical materialism than suggesting that all life participates in some sort of animated life force. From this perspective, there are no two substances and, as Loyal Rue suggests in his recent book, "nature is all you need."[42] This is not a collapse into materialism, but a navigation through the two errors of reductive materialism (or reductive idealism) and radical dualism (two-substances of mind/matter or matter/energy). From the emergent perspective, all of life is continuously opening up toward new ways of becoming. Mind emerges from brain and cannot be reduced to brain, ideas emerge from matter and return to shape that matter, and neither can be reduced to the other. This is a type of radical immanence.

Such a notion of immanence places humans on the same plane as all other life-forms. We are meaning-making creatures or as Deacon notes, a "symbolic species."[43] Much more will be said regarding this in chapters 4 and 5, however here I would like to note that rethinking humans in such a way means that human consciousness and human imagination is not the only possible version of sentience, perspective, or concern. Our symbolic consciousness is an emergent feature that finds resonance in other creatures on the planet: primates, dolphins, elephants, and perhaps even cats, dogs, and all forms of experiential creatures. This means that humans can never exhaust the multiperspectival evolving planetary community regardless of how much we peer into reality with our scientific technologies. Our perspective will always be the *human* perspective of the universe, the *human* perspective of the whale, and the *human* perspective of the ecosystem. Finally, such a perspective also means that there is a possibility for the emergence of something new beyond our current level of consciousness in the planetary future. We are in the process, then, of evolving beyond ourselves.

Again, Deleuze and Guattari have similar notions of these ideas in their understandings of bodies and subjectivities as assemblages, and human assemblages as always making meaning through reimagining the world in "lines of flight" that deterritorialize and reterritorialize our becoming assemblages.[44] As assemblages humans are made up of the past of evolutionary becoming (cells, molecules, tissues, and evolutionary adaptations that lead to upright postures, among other things), present ecological communities (water, recycled atoms, air, and food of other plants and animals that become our tissues), and our future becoming (desires, imaginations, ecosystems, and communities that provide our contexts and meanings). The subject then, is always a plurisingular event of open becoming. As Nancy notes, "Being cannot be anything but being-with-one another, circulating in the with and as the with of this singularly plural coexistence."[45] As humans, we ought to think of as many lines of flight that will help us to deterritorialize (that is, break out of the constricted, habitual confines that keep us locked into foundational ways of thinking and becoming) and reterritorialize (that is, reassemble in new ways that hopefully promote planetary flourishing rather than stagnation). In this way, we play a part in the self-organizing nature of the becoming planet.

The theory of emergence suggests a way of thinking about newness that pays deep attention to context without assuming that we are reduced to that context. In other words, new ways of becoming are never ex nihilo, but neither are we bound and reduced to the past. Emergent newness is based on autopoesis and the idea that "something more emerges from nothing but." As Deacon notes, "A complete theory of the world that includes us, and our experience of the world, must make sense of the way that we are shaped by and emerge from such specific absences. What is absent matters."[46] Absence is what allows for organization and possibly the new spaces for creative novelty to emerge from what seems to be chaos. This will have some interesting overlap with the need for agnosticism, apophasis, and unknowing, to be discussed in the next chapter. Let me note that Deacon draws on emergence from not just the early emergentists but also from Lao Tzu and Taoism. In other words, the ways in which absence is dealt with in the nonsubstantive thought of Taoism helps to shape Deacon's own contemporary apparatus for understanding the becoming world. Such a world is, as I mentioned already, marked by autopoesis.

Autopoesis means "self-organization." This is simply the idea that there need be no external organizer of life, such as found in arguments that life is designed by God or matter is designed by Platonic forms. Molecules of water come together because there is a fit between two hydrogen molecules and an oxygen molecule. Forces either attract or repel. There is no reason to add on top of that some ultimate goal or design, though Deacon and other emergentists do talk about the end-directed life and the emergence of teleodynamics.[47] Another example often used by emergentists to describe autopoesis is the way in which many water molecules come together to form a body of water out of which emerges the phenomenon of buoyancy. Buoyancy doesn't happen with just one water molecule, but takes many coming together. This buoyancy is "something more from nothing but." In a similar way, or so emergence theorists might suggest, mind, ideas, and imagination emerge from neural connections in the brain. And, I would argue, mind, ideas, and imagination return to affect the world around us.

Again, Delezue and Guattari have an understanding of an open evolving future that needs no principle of ultimate organization. For them, life could unfold on a "thousand plateaus." Further, we are always in the process of evolving beyond our human present. We are becoming with plants, other animals, minerals, and technologies.[48] Much more will be said about this in chapter 5.

Through our understandings of evolution and ecology, we are beginning to break down species barriers and recognize our radical dependence and emergence from the rest of the natural world. We are more like assemblages or flows of culture, history, biology, and energy than we are distinct individuals or species. Our identities are neither formed from divine commandment nor from some natural laws set forth from the beginning of the universe that would dictate the emergence of *Homo sapiens sapiens*. Likewise, our futures are not sealed off and secured in some transcendent teleology: rather we are emergent entities, and the future is open to many different ways of evolution. Deacon sums up well the implications of this type of emergent reality for our identity. He writes: "I am not the same I. On the one hand, I have somehow lost the solidity that I once took for granted, me-the-physical-body is no longer so certain; and yet on the other hand my uncertainty about my place in the world, the place of meaning and value in the scheme of things, seems more assured with the realization that I may be more like the hole at the wheel's hub than the rim of the wheel itself."[49]

Again, as Deleuze and Guattari suggest, the rhizome rather than the root tree (arboreal) should become our image for ontology. A rhizome expands in many directions and sends shoots (or lines of flight) in many different directions. A rhizome has no detectable center, origin, or taproot, but is rather in the process of expanding in multiple directions. It, like Deacon's emergent "I," has no center, which is what allows for its unique way of becoming. So it is with the life we find ourselves in: the possibilities for the future are emergent and multiple, there is no one, right way. Finally, our understanding of cosmology suggests that the universe is expanding in all directions. It is not closed off but expansive. Nonequilibrium thermodynamics also suggests that our universe itself may not be a closed system and hence entropy may not apply at the universal level.

The philosophies and sciences of immanence discussed thus far, from Bruno to Spinoza to Bergson and the emergentists and into the radical immanent theories of Deleuze and Guattari, all have some sort of influence from the influx of Eastern and indigenous thinking through the era of European colonization from the seventeenth to nineteenth centuries. Just as modern science contains within itself a substantial notion of metaphysics, anthropology, and ontology drawn from the Greek, Roman, and Christian heritage, so-called postmodern sciences are not devoid of religious influence. The influx of many different world religious traditions into Western academic education in the nineteenth and twentieth centuries is, I argue, part and parcel to the development of new scientific ways of understanding the worlds around us. Focus on nondualism (Einstein's relationship of matter and energy and quantum realities), process (evolution), and nonreductive sciences has surely been influenced by the influx of traditional ecological knowledge, Buddhism, Hinduism, and other traditions that focus more on relationship and process than essence and substance. In fact, these religions have been romanticized as being more ecological by Western environmentalists because Western notions of a romantic return to nature in a postindustrialized world were developed at the same time scientists and the thinkers of the West began to take seriously philosophies of Eastern and indigenous origins. In fact, as Said argued so well in *Orientalism*, there has never been a West without an East, and in reality these have always codefined one another.[50] The point here is not to suggest that substantive-influenced metaphysics and sciences are more or less natural or correct than nonsubstantive influenced metaphysics and sciences, but rather to see what types of worlds their corresponding regimes of truth help to cocreate. One regime I have already discussed is the Merchant-style death of nature or the industrialization and instrumentalization of the world for human purposes. The other, nonsubstantive regimes bring about the death of nature as something that is, has been, or ever will be something separate from humans, technology, and culture. It is toward this regime that I would now like to again focus the reader's attention.

NONEQUILIBRIUM THERMODYNAMICS AND OPEN EVOLVING SYSTEMS

> Preadaptations unstateable in advance, intersections between partially open systems of multiple kinds, and novel capacities for self-organization within a system triggered by infusions from elsewhere periodically operate in and upon each other, generating turns in time out of which a new equilibrium emerges, transcending our ability to articulate it in advance.
> WILLIAM CONNOLLY, *A World of Becoming*

The world of cosmology and physics has become increasingly more and more queer.

Einsteinien and post-Einsteinien nonsubstantive physics suggest that matter and energy are not separate, but rather matter and the things around us are collections of energy in space-time folds. This means that matter is always already internally energized or "alive." Furthermore, quantum and subquantum realities challenge the idea that all of life can be reduced to some basic, substantial level: below the atomic are the subatomic particles, below them are the quarks, neutrinos, and smaller and smaller quanta of reality to the point that nonsubstantive physics suggests there is no bottom to reality. This reality, arrived at through a different process and in no way the same as concepts found in ancient religious traditions, nevertheless makes more sense out of notions such as dependent coarising in Buddhism or the concept of Indra's net than it does of ensouled matter moving toward some sort of ultimate telos. These physics provide a different type of truth regime from that of a reductive physics tied to mechanical notions of reality, which literally employed physicists in the military industrial complex in the early half of the twentieth century. During that time, "very quickly, philosophical inquiry or open-ended speculation of the kind that Bohr, Einstein, Heisenberg, and Schrödinger had considered a prerequisite for serious work on quantum theory got shunted aside. 'Shut up and calculate' became the new rallying cry."[51] The *Gedanken* experiments of the earlier physics of Einstein and Bohr gave way, then, to practical, mechanical problems of industry and military. Such a narrowing meant that ultimate questions were all but forgotten. In his book, *How the Hippies Saved Physics*, David Kaiser argues that it was the generation of physicists during the sixties and seventies who, influenced by LSD and Eastern mysticism, began looking at the strangeness of the quantum world once again. Dark matter, theories of the multiverse, entanglement, and other ideas became the topics of retreats at the Esalen Institute in northern California. It is due to these thought experiments that the philosophical and speculative side of physics began to resurface and cosmologists once again began asking questions about the nature of the universe. Figures such as Capra and his *Tao of Physics* were some of the pioneers in reigniting the more theoretical approaches to physics and its relevance for how we live our lives. Eric Schneider and Dorian Sagan were influenced by this resurgence, and in their book, *Into the Cool: Energy Flow, Thermodynamics and Life*, they argue that entropy only holds for closed systems and that nothing in our world is a closed system (with the possible exception of the universe itself). Thus we live in a nonequilibrium state: a state of constant flow and change and one in which equilibrium equals death.[52] Barad draws on these theoretical physicists to develop her understanding of agential realism and suggest that matter is discursive, all the way down. In about four hundred pages she sides with Niels Bohr and indeterminancy over Heisenberg's uncertainty. In other words, it is not just that our epistemology is somehow unable to grasp the quantum world or reality in full; it is that the very way we look at the world shapes reality into different ways of becoming. However, this is not complete human construction of the rest of the natural world. She writes, "We are responsible for the world within which we live, not because it is an arbitrary construction of our choosing, but because it is sedimented out of particular practices that we have a role in shaping."[53] In other words, like emergent theorists and some of the post thinkers that have been and will be explored throughout this text, Barad suggests that the world is material-ideal or always and already epistemology-ontology, nature-culture. In the end, she suggests that "space and time are phenomenal, that is, they are intra-actively produced in the making of phenomena; neither space nor time exist as determinate givens outside of phenomena."⁵⁴ What this means is that we are radically contextual, always evolving phenomena, and that we are cocreated through our evolving ecosocial contexts and return to create those contexts. We may actually exist within one of many universes in a multiverse. The Hadron Collider near Geneva will test some of the theories of multiverse in a way that could make our knowledge of the universe shift as much as Copernicus's and Galileo's did in their time. Furthermore, even within our own universe, we are a part of a 13.7-billion-year process of ongoing cosmic expansion. To think that our human conscious experience could exhaust the realities of this 13.7-billion-year process is at least hubris if not completely arrogant. In this radical context, where multiple possibilities can lead to multiple realities, objectivity is always embodied and response-able. "Objectivity is a matter of accountability for what materializes, for what comes to be. It matters which cuts are enacted; different cuts enact different materialized becomings."⁵⁵

Surely these postmodern trajectories in the sciences suggest challenges for our meaning-making practices in the world. They suggest new ways of relating to the rest of the natural world, to other animals, and to the expanding cosmos. They do not dictate what or how we ought to become, but they do challenge quite a bit of our axial age methods of making meaning out of the rest of the natural world. Such ways of thinking, without strict borders, with permeable concepts, and with a strong understanding of the coconstruction of concepts and identities, are also the central component of queer theory. In fact, in the past decade queer theory has finally begun to challenge even the gender dimorphism and heterosexist assumptions in scientific explanations of the world.⁵⁶ It is to queer science and queer nature that I now turn. Such an understanding provides us with fertile ground from which to imagine planetary identities (chapter 5), communities (chapter 6), and futures beyond the present (chapter 7).

Queering Nature

> Specifically, the task of a queer ecology is to probe the intersections of sex and nature with an eye to developing a sexual politics that more clearly includes considerations of the natural world and its biosocial constitution and an environmental politics that demonstrates an understanding of the ways in which sexual relations organize and influence both the material world of nature and our perceptions, experiences, and constitutions of that world.
> CATRIONA MORTIMER-SANDILANDS AND BRUCE ERICKSON,
> *Queer Ecologies*

Though the queering of identities will be further discussed in chapter 4, here I want to outline the ways in which queer theory has something to provide even in conversations about nature. Science, to almost no one's surprise, has been influenced by sexism, racism, and even heterosexism. From the seizure of medical practices from women and indigenous peoples to eugenics to reading heterosexuality as the norm or natural for the rest of the animal world, science is always already influenced by the thought habits of those doing science. This is nothing new. Feminist philosophers of science such as Sandra Harding, Donna Haraway, Karen Warren, Carolyn Merchant, and many others have been saying this for years. What is new is the way in which queer theory suggests that there may indeed be no basis for any one category or analysis of evolving bodies in nature. In other words, it is not just that alternative identities and realities have been suppressed by racist, heterosexist, and patriarchal assumptions, but that identity itself, categorization itself, and models themselves may be arbitrary. As Barad suggests, knowing is not about representation, but is rather "an ontological performance of the world in its ongoing articulation."[57] Queer theory, like many of the other theories discussed thus far, transgresses any rigid boundaries. In this chapter I introduce three concepts from queer theory that are particularly important for ideas of nature: performativity, abjection, and the formation of identities in and through difference. More ideas from the area of queer theory will be discussed in subsequent chapters, but these three concepts bear directly on the topic of nature.

In *Bodies That Matter*, Judith Butler describes a process of identity formation that resists any sort of essential or constructed reading. In other words, for Butler, identity (and, more specifically for her, sexual and gender identity) is formed in a contextual process that includes histories and biologies that become habits for ways of being in the world. It is not that we construct our identities ex nihilo, as constructivists would have it, nor is it that our identities are predetermined by biology or religion; rather, our identities are performed. Through cultural-religious rituals and medical interventions, the ideas of gender and sexual dimorphism are imposed upon bodies from womb to tomb.[58] Though Butler does not draw much from the sciences to come to this conclusion, many other scientists have begun to take her Foucault-influenced theory and have begun to trace the ways in which heteronormativity has created a regime of truth that forces gender and sexual dimorphism on the world. Molecular biologist Joan Roughgarden, for instance, challenges the heterosexism inherent in evolutionary biology and the concept of sexual selection and then reveals the multiple sexual identities that exist and persist in the rest of the natural world.[59] From third and fourth genders, intersexuality, hermaphrodites, asexual reproduction, organisms that shift genders and sexes throughout life, and more, the idea that heterosexuality is the norm through which nature reproduces itself is challenged. Again, there is an inherent problem in trying to trouble these identities by searching for these identities in nature. However, the concept of performativity suggests that there is nothing natural about nature at all.

Performativity does not mean willful performance of a scripted role in a play; rather, it means that certain habits and systems, biological and cultural, provide ways of becoming that one must subject to in order to recognize his agency at all. The process of subjection—to these habits and roles—is what provides the agency to perform in the first place. Far from demonizing roles, concepts, and distinctions, we need them in order to be subjects at all. However, as Marcella Althaus-Reid suggests, every version of a performance is a per/version (another version).[60] It is in these per/versions of performing identities that they begin to shift, change, and transform. As Foucault, Butler, and most other queer theorists suggest, the whole notion of gender or sexual identity is something that is fluid and changes over time. Such notions of performativity may be a better way of thinking about nature in general.

Nature is not a stable entity, as any scientist will confirm. Concepts of nature are even less stable, as any philosophically astute reader will confirm. Why, then, do we continually resort to notions of what is natural or unnatural in order to justify arguments? Instead, we might think of the evolving planetary process as a performance that draws on the habits of nature. In the repetition of these habits emerge differences leading to evolutionary shifts toward new ways of becoming. From this perspective the laws of nature are much more like pathways, scripts, or habits that get performed, and in every performance there is some per/version, leading to new possible ways for the future of nature naturing, including humans, culture, and technology here. Thus performativity is a metaphor for the becoming process of reality. As such, a concept of nature will always emerge, but we must take responsibility for that concept of nature and what it leaves out. "Intelligibility is a mater of differential responsiveness, as performatively articulated and accountable, to what matters."[61] From the perspective of queer theory, any snapshot or concept of nature at a given time is always leaving something out: the deferred or the abject. There is in our relationship to the world between thoughts and material (though always coconstitutive of one another) an "irreducible difference."[62] It is this abjection, this differance, that provides for the possibility of new, creative ways of becoming.

Abjection, among other things, is central to the process of identity and concept formation. It is the remainder that is left over when we identify something as this *and not that*. It is the *not that* that is central to the very process of identifying what *this* is. We can think of it in terms of identifying something as natural, which always implies the unnatural. Or identifying the good, which always implies the bad. This process is not problematic in itself, but the tendency in identity and concept formation is to deny the very existence of the other, which all critical theory (including feminist, race, and queer) has brought to light. When the process of identity formation denies the existence of the other that is central to its very formation, conceptual and actual violence often results.

In terms of nature, environmentalists often assume that nature is all that is not human. The human, cultural, and technological are then the denied abject that return to haunt the very definition of nature. This results in a bind. It is the natural that is good and ought to be preserved, but it leaves no place for the abject, which is the very thing troubling the natural. For too long, technology, culture, and human beings have been written out of nature by environmentalists, religions, and many sciences. The failure to include the abject (human) results in the very failure of environmentalism to attain its goal: the conservation, preservation, or sustainability of the natural world. This human exceptionalism will be addressed in chapter 7, the final chapter of the book, but here I note that any definition of nature that does not include always and already humans, imagination, religions, cultures, technology, and ideas will inevitably fail. Of course, in any new conceptualization of nature there will always be an abject that future manifestations must return to address. This is what Latour hints at in his idea of the collective that must continuously deal with that which is left out of any of its given manifestations.[63] The point of recognizing the abject is not that somehow there will be a transcendent point at which all life is included, but rather to continuously trouble the permeable edges of our thought toward ever new ways of becoming that seek to bring the claims of the abjected to our attention. Such a focus on process helps us to come to terms with the fact that we are always already coming into being in a collective process of repetition of differences.

As mentioned earlier, the formation of identity has often been thought of as a process where like chooses like. Since at least the time of Aristotle, knowledge is based upon human reason identifying reasonable laws in nature or, in a Platonic sense, recognizing the eternal essence of a given thing as reflected by some unchanging Form. Queer theory, and poststructural thought in general, suggests that we only know in and through our differences. Oneself, in a given context, only recognizes that she is a self through recognizing that she is not an other: a table, a bird, or another human. Postcolonial theory understands this as the process of being identified in the interstice—the exchange between self and other.[64] The recognition that our own identities are only formed through interactions with *differends* is a bit like understanding the Buddhist no-self concept of self. In other words, we are contextual beings through and through, and it is only through our interactions with human and earth others that we ourselves are formed.

The entire ongoing process of planetary evolution can be seen as this type of contextual process. Ecosystems are only as good as they are open to their ecotonal edges through which information is exchanged. The ecotone is the edge of an ecosystem that literally prevents an ecosystem from becoming closed in on itself and perishing, much like the permeable nature of cells or our bodies that keep us open to the very possibility for future becoming. As Sagan suggests, equilibrium is death.[65] This means that we are literally always becoming plant, mineral, and animal with the evolving contexts of the planetary community. It also means that emergent technologies will change the future of planetary becoming, close off some past possibilities for becoming, and open up new ways of becoming. There is no return to some sort of pure nature, for there is no nature to which we can return. This does not mean anything goes, but rather it means we must have a different way of thinking about ourselves in relation to human and earth others (the subject of chapters 5–7).

Again, both science and religion have been guilty of the logic of globalization: of spreading one meaning-making practice, one understanding of nature and reality, over the face of the entire globe. This is just as true for the process of the failed green revolution in agriculture as it is for the efforts of early colonizers to the Americas and their attempts to spread Christianity. If truth is just out there, waiting to be grasped, and if that truth is singular, what choice does one have but to educate, civilize, and cultivate awareness of this truth in any other he might come into contact with? This, at least, is the logic of globalization. From a planetary perspective, however, truth is seen as the coconstruction of truth regimes. Our understandings of the world and the technologies of those understandings begin to create those worlds that we are persuaded most toward.[66] In other words, one of the reasons modern science became so pervasive is that its truth regime—including the medical, communication, and transportation technologies derived from its way of understanding—is quite persuasive. It gives us results; it gives us things. However, at no small cost: atomic bombs, environmental ills, species extinction, global climate change, and gross economic inequities are just a few of those costs. What I am suggesting here is that one can live truthfully within this world of modern science, but there is always a cost. This is the case of any meaning-making system and its truth regimes. One can live truthfully from within the truths of traditional ecological knowledge and in relative isolation from the forces of globalization and development as well—again, not without some costs. Every truth regime, and its corresponding habits for becoming in the world, has benefits and costs, and this is what it means to understand truth from a pragmatic perspective. From a planetary perspective, the question is not which truth regime is really real, but rather toward which truth regimes do we want to live? Given the costs of the contemporary truth regime of the globalization of free-market capitalism and its modern scientific technologies, I would argue we need new ways of becoming into the future that respect the multiperspectival reality of the becoming planetary community. We need to begin imagining with the whole planetary community in order to develop new ways of becoming into the future. These new ways do not need to be singular, as the wider planetary community has largely thrived on biodiversity, and human communities on biocultural-historical diversity. Rather, the point is that through this rethinking human becomings are thought back into the rest of the natural world and that this nature is understood as always already a political process. Planetary politics, then, will extend the category of the political to include the rest of life on the planet and also place critical focus on any understanding of nature that becomes naturalized.[67] Before we get to planetary politics (chapters 5–7), we must first destabilize foundational understandings of religion (chapter 3) and identity (chapter 4). Without destabilizing all three of these sources—nature, religion/meaning making, and self-identity—the inertia of the false foundations they can create pulls us back toward their centers, and worlds are monologically recreated in their images.

3 DESTABILIZING RELIGION

The Death of God, a Viable Agnosticism, and the Embrace of Polydoxy

> If ideation is electrochemistry, electrochemistry grounds, rather than undermines, all ideation. Therefore, to eliminate one ideation (that has its electrochemical grounds) in favor of another cannot be grounded in physics.
> IAN HAMILTON GRANT, Philosophies of Nature After Schelling

Inasmuch as there has been a destabilization of the concept of nature complete with a geneaology of scientists who understand nature as something other than stable, mechanical, or anything that can be contained, it is also true with religion or meaning making. Furthermore, inasmuch as religion has helped to destabilize nature, so science has helped to destabilize foundational understandings of religion, meaning, and value. In other words, religion and science are always already involved in codefining one another. This chapter will explore the ways in which religion can be rethought if grounded in these evolving ecosocial contexts. Religion, rather than being above the fray, is always focusing our attention back to our current contexts, sometimes in confirming and cataphatic ways and other times in iconoclastic and apophatic ways. As Bruno Latour notes, "religion . . . does everything to constantly redirect attention by systematically breaking the will to go away, to ignore, to be indifferent, blasé, bored."[1]

These contexts I describe are emergent and as such always involve unknowing at the edges of our understanding. We cannot continue to project certainty where our knowledge does not tread. In fact, certainty is only possible, in a planetary perspective, when we narrow the many possible ways of becoming down to a single possibility. If knowledge is only a regime for organizing the world, then representationalism is out the window, and we must evaluate knowledge based upon its ethical outcomes. As such, certainty is always problematic in that it suggests a logic of conforming all other possibilities to a single, dominant one. This process ends up causing conceptual and physical violence. Thus this chapter is an attempt to ground this way of understanding religion in the apophatic, negative, iconoclastic, and deconstructive strands that can be found in many of the world's religious and philosophical traditions. Such ways of thinking, far from being a phenomenon that is recent and limited to "post" ways of thinking in the twentieth century, are crucial for helping us to understand ourselves as evolving, planetary creatures. They help us to understand the phenomenon of being open, evolving meaning-making creatures (chapter 5) in a world that is constantly becoming (chapters 6 and 7). As such, this chapter calls for regrounding religions in a planetary context. One such way of doing so is to think of our meaning-making practices (including religions) as imaginative lines of flight.[2] These lines of flight into an unknown future, rather than sealing off the past and future into a foundational present, keep our meanings porous and open to unknown human and earth others while attempting to cocreate contexts that facilitate creative changes toward the flourishing of the future planetary community. As such the *place* of meaning making, including theology, is at the ecotonal edges between identities. It is the process that facilitates exchange between self and other, between humans and the rest of the natural world, and between understandings of nature. Such openness to others in terms of our meaning-making practices is only possible with a certain amount of apophasis in our meaning-full coconstructions. I begin this chapter then with a discussion of the relevance of "the death of God" and other apophatic concepts in some of the world's religions. Again, my aim here is not to provide a linear origin for contemporary deconstructive methods, but rather to provide multiple historical sources for contemporary contexts of making meaning. These various groundings help provide us with tools of persuasion toward understanding planetary meaning-making practices as multiple lines of flight and suggest that we ought to embrace a bit of polydoxy rather than argue over the illusory concept of orthodoxy.

Unknowing at the Edges: Making Meaning in the Dark

At the edges of our meaning-making projects we must admit, and learn to love as part of our meaning-making biohistorical nature, darkness, the unknown, and mystery. It is from these dark matters, these unknowns, and these mysteries that the creative-destructive processes of life become reenergized and renewed. Once these dark spaces are filled with the projected light of certainty and the living stream of life is damned up, equilibrium in nature and culture will lead eventually to cold death.[3] A nonequilibrium and thus viable theological understanding of the world should, I argue, also be agnostic. We need to relearn the virtue of unknowing to get beyond our quest for certainties or even, as David Byrne suggests with the title of one of his albums, and as philosopher John Llewellyn might agree, "stop making sense."[4] As we begin to recognize that our meaning-making practices find their groundings in evolving contexts that shade off into mysteries, and our sense making loosens its control at the edges of our understandings, perhaps we will begin to open up toward the evolving planetary contexts in which we exist and begin coconstructing with many earth others a more inclusive, participatory planetary democracy.

The problem with certainty, as Catherine Keller and many others have pointed out, is that it creates a conceptual violence that leads to violence toward other bodies.[5] Certainty blocks our imaginative thought projects by attempting to contain that which is abject, other, uncertain, or unknown. Such is the problem with global ways of thinking in the world. Globalization is marked by the reification of life into certain, globalized, and capitalized ways of becoming. Our imaginative energies, rather than being freed up to think of new ways of becoming with the planetary community, get placed into preconceived ways of being in the world. In other words, our imaginative energies become what Martin Heidegger calls "standing reserve" for the continued globalization of free-market capitalism. In fact, as Philip Goodchild argues, humanism has run aground because its very existence depends on the mastery of nature via science, control of nature via technology, and liberation from nature via capitalism. In other words, humanism depends upon making all other life into standing reserve for the human being or recreating the world in the image of human becoming.[6] Rather than imagining open possibilities for becoming, trying to solve issues such as poverty, economic disparity, global climate change, and other ecological and social ills, all imagination is reduced to how to deal with these problems from the globalizing market perspective. The one-fifth perspective literally creates the world in its own image, leading to the destruction of many bodies and the closing off of open, emergent possibilities. There simply are not enough materials to actualize all perspectives for planetary becoming, which is the promise of free-market capitalism. Again, Goodchild argues: "The modern quest for wealth—the increasing domination of the natural world—and freedom—the separation from natural constraint and social obligation—are illusions, impossible ideals born of representation and abstraction, projections of an idealized condition in which humanity cannot survive or flourish."[7]

What happens as more and more capital is created is that there is a negative correlation in imaginative thought exercises. The more and more capital that accumulates behind a specific way of becoming (even it is for the benefit of fewer and fewer planetary beings) means that that way of becoming has the resources to be actualized with less and less thought and less and less effort. The world literally becomes reified into a certain way of becoming. Once more and more resources get caught up in a single line of flight, it becomes easier and easier to capitulate to that line of flight because doing otherwise takes a lot of imaginative capital, and labor power: which is precisely what is worn away through reification. In light of this situation the humanities, and religious studies, in particular, are *crucial* as we begin to move toward planetary identities. More will be said on this in chapter 7, but the economization of higher education is one of the great tragedies of our contemporary era. Such economization means that our thoughts and imaginations are forced into becoming in the image of the global market, the very same global market helping to create problems such as gross economic inequity and global climate change. One wonders here, along with Audrey Lorde, "can the master's tools be used to dismantle the master's house?"[8] On the one hand, we cannot escape the process of globalization: it is part of our contemporary contextuality. On the other hand, we must not allow such globalization to close off future, emergent per/versions for becoming. This would amount to reifying life into the image of the past or that which is realizable through capital.

The problem with hitching our imaginative powers to the realizable realm alone is that we become insensitive to the agnostic nature of our realities. In other words, the more we can create worlds in our own images, the easier it is to reify identities and concepts through opting for foundationalism or circularity in our thinking. In doing so, these foundational identities and concepts create violence as they background the evolving and open processes involving the rest of the planetary community. As Zygmunt Bauman suggested in his book *Globalization*, those with the capital to do so become more and more confirmed in their knowledge as knowledge of reality and in their identities as *true* identities as they create more and more of the world in their own image.[9] There becomes less and less need to recognize the limits of human abilities to grasp reality in concepts and language the more and more economic power one has. Further, such power is associated with freedom, and this becomes the goal for "developing" countries. Even the very language of "freedom" then becomes tied to understandings of economic freedom to assert one's isolated identity over and against any other apart from any consideration of contextuality.

Thus it is not just in religion that we need a bit of apophasis, but in all of our ways of meaning making. There are many resources within the world's religions that support apophasis, but here I will discuss three: the death of God, the Buddhist concept of emptiness and form, and the image of the trickster in many indigenous traditions. Each of these resources offers us ways to make meaning that does not close us off from the ever evolving planetary contexts of which we are a part; rather they allow us to remain open to emergent, new ways of becoming with multiple earth others.

The Death of God

> Without the abandonment of accredited certainties we remain inattentive to the advent of the strange; we ignore those moments of sacred enfleshment when the future erupts through the continuum of time.
> RICHARD KEARNEY, *Anatheism*

It was, of course, Nietzsche who first and most famously proclaimed the death of God. What Nietzsche and subsequent death-of-God thinkers were after was not the end of religion, but the end of a transcendent referent from which the whole world could be recapitulated or understood. In other words, this claim is akin to claiming that there is no Archimedean standpoint, no objective person, form, or power that offers the view of totality. The Omni-God (omniscient, omnipotent, omnipresent) is nothing more than a projection of our desires for certainty where none exists. Nietzsche is not the first to notice this. There were many apophatic and negative traditions within Western religions that understood the danger of claims of omnipotence. The mystical traditions in particular had a good relationship with the uncertainty at the edges of our knowledge. Keller has done an excellent job of highlighting the apophatic tradition of Pseudo-Dionysius the Areopagite and Nicolas of Cusa and their relevance for poststructural, deconstructionist, and post thinking in theology in general.[10] It is important to remember here as well that Pseudo-Dionysius's thought was important for the development of Cusa and then that of Giordano Bruno.

Such recognition of the unknowing at the edges of our own knowing leads to understanding agnosticism as part and parcel of our very process of meaning making. In order to acknowledge the process by which we cocreate with multiple earth others, we must maintain such agnosticism at the edges of what we know. Just as the person of Jesus of Nazareth could not have imagined the world of iPhones, airplane travel, and remote consumption through the Internet, so now we cannot imagine the possibilities for becoming in the world two thousand years from now. What happens when we try to impose certainty is the imposition of sameness over all times and places, which is the very same process that takes place in colonization.[11] This apophatic tradition is then a built-in deconstructive element to theological projections that allows for otherness and emergent newness in the process of planetary becoming. In other words, these negative traditions remind us that in the end our knowledge is partial and that these partial knowings produce effects; what we work for and believe in is but one trajectory and constantly engaging the others within and outside our own constructs is necessary if we are to thrive in open, evolving contexts.

Emptiness and Form: Dependent Coarising

Another source from "traditional" world religions that recognizes the need for agnosticism and apophasis is the Buddhist concept of dependent coarising or *pratītyasamutpāda*. Again, it is such notions of nonsubstantive metaphysics as found in Buddhism that lead to what some have called the postmodern sciences. Thus meaning-making practices always and already involve worldviews that combine both religious and scientific information. Here, however, I want to discuss how dependent coarising and the stress on emptiness and form help us to develop a bit of agnosticism at the edges of our knowing.

The idea of dependent coarising in general means that all life, present, past, and future, is in a real sense interconnected. One simple and popular example of this can be understood by simply staring at this page. What is this page made of? At first you will probably think: paper. However soon you will have to acknowledge trees, and the loggers that cut the trees, and the sun, dirt, water, and air that go into making the tree, and eventually into the elements that make up all life, and finally all the way back to the big bang 13.7 billion years ago. In other words, our concepts and our bodily forms are in a real sense empty or nonenduring. We are made up of the air, plant, animal, and mineral life around us and will become life for others in the future. In a very real sense the conceptual and physical boundaries we use to mark off self from other, human from animal, plant from animal, and life from death are arbitrary truth regimes. Thus we ought not hold on to our concepts as if they represent reality. Such grasping is the source of pain and suffering that the Buddhas and all bodhisattvas seek to liberate us from.

This emptiness of form that dependent coarising implies suggests that we are assemblages in the Deleuzian sense. We are made up of plant, animal, mineral assemblages and we are evolving with these other life-forms toward an unknown future pregnant with wild possibilities for transgressing boundaries of self/other, human/nonhuman, animal/plant, and organic/machine. To take a slice of reality and impose it on the rest of the natural world then is to reify a specific moment in time and to force the rest of the planetary becoming into the image of that same moment in time. Again, this is to make what Heidegger calls standing reserve out of the rest of the becoming planetary community. Such a desire to recreate the now in one's own image is, according to Buddhism, the source of suffering. Thus recognizing the unknowning and unraveling of our concepts and identities at their edges might open us onto planetary others in a continual process of exchange and cocreation. This is also the role of the third source for agnostic meaning-making practices I want to discuss here, the trickster figure.

Tricksters as Creative Destroyers

In short, trickster is a boundary crosser. Every group has its edge, its sense of in and out, and trickster is always there, at the gates of the city and the gates of life, making sure there is commerce. He also attends to the internal boundaries by which groups articulate their social life. We constantly distinguish—right and wrong, sacred and profane, clean and dirty, male and female, young and old, living and dead—and in every case trickster will cross the line and confuse the distinction.[12]

The trickster figure in many indigenous communities is akin to the creative destruction of the dancing Shiva from Vedic traditions. The ring of fire surrounding the Shiva is meant to burn through reified worldviews precisely because impermanence rather than endurance of concepts, substance or forms is what marks reality. Furthermore, Shiva stands on the ego, suggesting that it is our egoistic impositions on the world, impositions of certainty, that actually create violence. Similarly, trickster figures are the go-betweens in many indigenous cultures that navigate between the porous boundaries of self/other, human/nonhuman, between different communities, between ancestors of the past and the present living community, even between the binary gender system of male and female, and between the indigenous tradition and postcolonial identities of the communities.[13] It is the recognition that every boundary/borderland is not so much a hermeneutically sealed wall as it is a porous cellular wall that exchanges information with others.[14] In such an exchange entities are transformed. In this transformation the present must be destroyed for a new present to emerge, for a self to continue living on as an other.

Such exchange and the trickster figures that oversee them are not seen as moral or immoral, or good or bad, but rather as necessary for life to continue into future present becomings. There is then recognition that at the edges of our knowing we must recognize that the present moment is situated between a past that clouds off into unknowing and a future that shades off into uncertainty. We ought then to leave these edges open to becoming unknowns rather than closing them off to the certainty of the present. In their recent book, *The Quest for the Historical Satan*, Miguel De La Torre and Albert Hernandez suggest that Satan historically played such a trickster figure within the monotheistic Christian West. Such an understanding today might help Satan become less of a bogeyman and more of a tool that might be used to trip up ossified structures and unjust institutions—or just lead to a little creative destruction when categories of good and evil become too rigid.[15]

Death of God and postdeath of God theologies, dependent coarising, and trickster figures all provide well-water for apophatic ways of thinking in our own meaning-making practices. Such apophatic ways of thinking also find resonance with emergent, and evolutionary understandings of the natural world, and will have deep implications for understandings of identities (chapter 4 and 5) and ethics (chapter 6 and 7). At this point in the book, however, I want to begin to tie some threads together from postunderstandings of both nature and religion that begin to locate us in our cotemporary, emergent biohistorical contexts of planetary becoming. Such emergent meaning-making practices insist that in order to continue into the future toward a thriving planetary community, the only viable option at the edges of our meaning will be unknowing or what I refer to as a viable agnosticism.

Emergent Meaning-Making Practices: Toward a Viable Agnosticism

> Emergence theories presuppose that the once-popular project of complete explanatory reduction—that is, explaining all phenomena in the natural world in terms of the objects and laws of physics—is finally impossible.
> PHILIP CLAYTON, *The Re-Emergence of Emergence*

Rather than rehearse the history of emergence here or rehash the discussion in the previous chapter, I want to analyze one way in which emergent understandings of nature are most relevant for religion and meaning-making practices in general.[16] Far from reducing religion and meaning making to mere projection, emergence theory provides us with a way that describes how ideas, imagination, and meaning-making practices matter to the world without reducing them to some materialistic glitch or creating some sort of metaphysical dualism. Emergence can provide a way to understand bottom-up and top-down causation without falling into idealism or materialism.[17] The heart of this is in the discussion of autopoesis (see chapter 2).

Remember, autopoesis is the self-organizing principle of ever evolving life. "When aggregates of material particles attain an appropriate level of organizational complexity, genuinely novel properties emerge in these complex systems.... Emergent properties are irreducible to, and unpredictable from, the lower-level phenomena from which they emerge."[18] Autopoesis then describes a context-dependent understanding of newness. These new emergent levels of reality are neither discovered nor placed into existence by some discontinuous, external force. They are neither reducible to the same nor are they created out of nothing. Rather, new things, much like new ideas, emerge from and move beyond specific biohistorical/natural-cultural contexts. These new emergents are not merely caused by, but they also have causal effects on the contexts from which they emerge.[19]

There is, with emergence theory, the beginnings of a discussion on how to understand top-down causation as well as bottom-up causation. By top-down I mean, "Higher-level entities causally affect their lower-level constituents."[20] Though many materialistic scientists might cringe at the suggestion, I would argue along with other nonreductive scientists that emergence theory might be a viable way forward through the Scylla of reductive monism and the Charybdis of Cartesian dualism. That is, I am arguing, as does the late theologian Gordon Kaufman, that our biohistorical projections have efficacy for what philosopher Donna Haraway describes as the natural-cultural worlds in which we live. Our mental events affect our brains, not just vice versa. Newness, then, comes at both the levels of biology and history. At times, our biohistorical projections—our God-talk, utopian ideals, etc.—can really affect the way the world becomes in such a way that new, complex relationships emerge. Granted, this newness does not always mean "good;" newness, as the amoral trickster might tell you, works both ways. The point is that both bottom-up and top-down causation (though neither should be confused with direct causation) are necessary if there is to be newness without merely more of the same. From a bottom-up only explanation, everything is explainable from the same laws of physics. From a top-down explanation, everything is ultimately explainable by some external force—whether a destiny set in the stars or an omni-God. Neither a top-down nor a bottom-up approach is enough to explain the evolving planetary realities that make up our daily lives. Without both, we are taken out of our biohistorical contexts.

Both the scientific method of reduction, and foundational understandings of theology have been guilty of cutting natural-cultural life off from its living source. Furthermore, both scientific and religious epistemologies have been guilty of ignoring Horkheimer and Adorno's caution in *Dialectic of Enlightenment*. In a world where "reasonable" is equated with "reality"—whether one takes a theological, scientific, or combined approach to reality—myths/foundations are constructed to fit all of reality into the human confines of reason. As evolving realities tend to be much more than human reason alone, and much life escapes reason, the result is a reification of a part of reality and a mistaking it for the whole of reality. The effect of this process is alienation and violence toward all others that don't fit into the most reasonable construction: in other words, the effect is madness. I'm not arguing that we should abandon reason for pleasure or imagination alone, but we should recognize the limits of reason in capturing reality. Such an acknowledgment may help to guide our meaning-making practices out of the trap of ethical anthropocentrism while acknowledging that our context as human beings, among other things, always already means that we cannot avoid epistemological anthropocentrism.[21] After all, as Robert Bellah points out (in over five hundred pages!), we should precisely look at ourselves as narrative creatures within an evolving, planetary context, which always means paying attention to our human-ness. "Human beings are narrative creatures. Narrativity . . . is at the heart of our identity."[22]

In what follows, I suggest that a viable system of knowledge and meaning making, a living one, ought to be one that recognizes the limits of reason (which always ends up in foundations or in circular argumentation) and opt rather for a dialogical method of infinite regress. Before I describe what I mean by an epistemology of infinite regress, I first want to offer a metaphor to help evoke a sense of the type of context out of which my own thinking emerges. This context is very much shaped by the fluid reality of globalization that I discuss throughout chapters of this book. Here, though, my attempt is to evoke a sense of the way in which we might be able to embody more this type of fluidity as it relates to our understandings of knowledge and the process of meaning making.[23] As such, we must examine briefly the space-time of our meaning-making projects. Forcing all of reality into the confines of reason imposes a linear space-time on the planet. Colonizing metaphors that move from dark to light, barbaric to civilized, and even notions of progress and development are forced onto the planetary community when that community and all bodies therein are reduced to the confines and technologies of reason or any other such "tunnel of time."[24] Thus the space-time of a viable, agnostic meaning-making process will have to challenge the linear narratives behind much of Western thought.[25]

Nothing quite marks the living experience of the late twentieth and early twenty-first centuries as aptly as "contextuality," the notion of existing in located positions. Such locatedness is marked by the time and existential experience of ambiguity, as William Connolly describes in the opening quote to this section. Identity politics, rights movements, liberation struggles, and other responses and reactions to colonization and communicative and economic globalization lead to an experiential acknowledgment of ourselves as located. However, they also spur us toward the end of thinking about the end of history. Once we see that we are many, of many stories, of many religions, of many identities, many truths, many values *and* we see that we are—as the Apollo 8 image "earthrise" and the problem of global climate change suggest—of one planet, we are at a loss for words that make sense of it all. Linear narratives and histories don't really capture the reality of our era. Rather, some form of reverberating contextuality comes to the fore. That is, we are male, female, heterosexual, homosexual, bi, trans, queer, black, white, brown, latino/a, American, Japanese, Kenyan, and more generally of specific descents, yet we are also animal, biological, planetary, and ecological, inextricably bound within and to the planet Earth. "We" are not really a "we" without some sort of context, but the text that we are with is never stable, and the "we" is always shifting with the text between greater and lesser degrees of inclusivity. What do we do with this paradox between the need to claim particular identities (though evolving) and affirm our planetary emergence and bodily existence (some sense of common context)? In a sense the religious traditions discussed earlier—the death of God, dependent coarising, and tricksters—all make much more sense out of these fluid interconnected identities than do metaphors of embodied souls that move through a linear progression or digression from birth to death to some sort of eternal salvation. Such time is only possible with the clear delineations between self and other assumed by substance-based metaphysics. However, time must be thought of in different ways if the self is defined contextually through relationships with others and without foundations in either nature or religion (see chapter 4).

Perhaps we could think of the time that takes place in this new space, which is the dash of all the *post-* thinking, as reverberation rather than as linear or cyclical. That is, from the present time we look toward the past, but only from our present context, which is also always defined by our hopes/dreams and visions of future becomings. Who is to say whether the voice is coming from the future or the past? If you have ever been sitting between two buildings, or in a canyon, and hear a noise that seems as if it is coming from in front of you but is really coming from behind you, then you will get what I mean by this experience. The present reverberates in much the same way between past and future, except the mathematics and physics of sound do not offer such precision in discernment when it comes to the realm of the present experience of past and future. Is what we interpret as history where we want to go? Is what we interpret as history rooted in where we have come from? Both versions of history, those relying on origins and ends, are placed into question.

In a reverberating present—marked by infinite regress in terms of origin, the context that gives rise and shape to the sound and openness as the sound travels outward and cannot be contained—our epistemologies and systems of meaning should be geared toward dialogue, toward the recognition of multiple perspectives. There is no solid past and future, but the becoming present decides how we interpret the past in accord with what visions we desire for the future. This is the bold reality of the dash in the *post-* without knowing where we are going we cannot project certainty into the past. This is the Deleuzian-Guattarian rhizomatic ontology that recognizes contextual ethical responses without justification in origins and ends. The reverb of re-verb-eration implies, then, re-iterations or the ability to respond. Response-able. Responsible. Responsibility for the knowledge that we are always already *with* text, and taking responsibility for that text, is the place where our thinking begins. This is the habitat provided by the viable agnostic meaning-making practices that I articulate in this book.

A Call for Planetary Religions: Our Contexts for Meaning Making

The claim that humans are radically in the world in no way negates the uniqueness and incredible emergence of history and culture from the ongoing process of continuing creation. "The gradual emergence of human culture, of human activities and projects—has been as indispensable as were the biological evolutionary advances that preceded our appearance on planet Earth."[26] This historicity is that which makes human beings unique in the grand scheme of life.[27] It is made all the more awesome by the fact that it happens in such a vast universe. Historicity can arguably be seen on this planet as accelerating the rate of change in continuous creation for better or worse. "The question is not whether humans have induced change in ecosystems, but whether they have inordinately accelerated or inhibited change and in ways that are deleterious, whether to humans specifically or to the terrestrial life forms in general."[28] History and culture are forcings of biological evolution that emerge out of biological evolution rather than exceptional to the evolutionary process. Furthermore, human uniqueness from this biohistorical understanding of the human being does nothing to negate the uniqueness of all other life-forms. From this position of biohistory, human beings can begin to speak theologically about the worlds in which we become.

As biohistorical creatures existing as part of and within the evolving living world, theology moves away from metaphysics and foundational claims about God, at least from the substance metaphysics that characterizes much of the history of Western and Christian thought.[29] If we are not merely *biology*, nor merely *history*, then life cannot be reduced to either term. Reductionism and idealism aimed at forcing all of life to capitulate to nature, reason, or revelation no longer has a place in theological constructions.[30] Rather, theology becomes conversational. This is what Kaufman sees as a move from "first-order" or polemical theology to "third-order" theology. Third order theology recognizes theology as imaginative construction and suggests that "we must now take control (so far as possible) of our theological activity and attempt deliberately to construct our concepts and images of God and the world; and then we must seek to see human existence in terms of these symbolic constructions."[31] Furthermore, such deliberate attempts follow the methods of performativity and strategic essentialisms that feminist and queer theorists have been articulating. In other words, we recognize that there is no pure space from which our meaning-making systems can be dictated, thus for which we are not responsible; rather, meaning-making practices are always power-filled, and as subjects within meaning-making systems we can attempt to create per/versions of practices that destabilize oppressive power dynamics. Following Foucault, truth here is always power-filled.[32] The realm of truth is resituated from a transcendent space of foundations to the evolving nature-cultures in which we live. "In this model, truth is never final or complete or unchanging: it develops and is transformed in unpredictable ways as the conversation proceeds."[33] Again, this represents a viable agnostic theological and epistemological understanding of the world.

As response-able biohistorical creatures, we can modify our symbolic structures to let more of life in, rather than seal ourselves off from life in our concepts. These new reconstructions are not, of course, ex nihilo. Ex nihilo newness, as Keller has so insightfully pointed out, has been a major source of colonial violence.[34] Kaufmann suggests, "It should not be supposed that the theologian creates the order into which he or she fits the multifarious features and dimensions of life simply *ex nihilo* . . . it is always based on the prior human constructive activity which produced and shaped the culture."[35] Even those considered radical revolutionaries cannot escape the biohistorical trajectories in which they exist. We can and do move and shape these trajectories, but we never move fully beyond them or into a no-where space outside them. In this sense, Western theology is an integral part of the understanding of those who have grown up in or been influenced by Western culture. "Western theology should be seen first of all, thus, as a part of the language and traditions which have shaped Western experience."[36] This makes God-talk all the more relevant in a culture that segregates religion to a space outside of the public realm.[37]

This understanding of theology as imaginative construction, as conversation, and as the ongoing process of modifying living symbols in no way negates the reality of what we might call the spiritual, cultural, or historical sides of life. This is often the critique waged against theological projects that fall within the trajectory of Feuerbach's insights: namely, that theology is thereby reduced to merely an individual projection and that God is not, therefore, real. However, what our biohistorical existence means is that the historical side, the cultural side, and the God side *are* as real as the bio and natural sides of the equation. "To regard God and the world as constructs with which we bring order and personal meaning into experience does not involve downgrading them in comparison with directly perceivable objects . . . indeed, we could not live and act and think without some such ordering principles or images."[38] Our concepts, our theologies, and our symbols are as much a part of our *real* life as those things we directly perceive. Indeed, this is one of the implications of bringing God and the world back together in an ongoing process of continuous creation. Despite some of Feuerbach's own negative impressions about God as projection, i.e., that projections of God are almost always unhealthy egoistic projections, many have suggested that Feuerbach can actually be taken in a different direction (and that Feuerbach himself was moving in that direction): namely, that theological projections are part of what it means to be human.[39] "In other words, the question is whether we can take the imaginative nature of God's origin not as an indicator of its non-reality, but rather as a sign of the poetic nature of God language and hence of the symbolic nature of God's reality."[40] In many ways this is what all post theologies are about. This is an attempt to *make real* the idea that whatever it means to be human includes meaning making and that the effects of meaning making—value, imaginations, hopes, dreams, and theologies—are indeed just as real as chemicals, neurons, genes, and atoms. What differentiates this type of theological projection from the type that Feuerbach critiques is precisely the agnostic, postfoundational character of the projections. These imaginings are not locked into the real/not real structure of Feuerbach's own modern mind, but rather comprise a unified reality. "The importance of Feuerbach's thesis that the imagination is the key to understanding how and why human beings are religious lies beyond the horizon of his own narrowly positivist epistemology and stolidly modernist temperament."[41] Furthermore, and perhaps a bit paradoxically, though Feuerbach's modified understanding of religious thinking/theology as projection/imagination may offer an evolutionary-emergent way of understanding religion as part of the natural-cultural world, it no longer need signify an evolutionary supercessionist understanding of the development of religious ideas. That is, the move from immanent and animistic to civilized and monotheistic ways of thinking about religion is no longer a part of a historical narrative of progression toward those in power now (i.e., Western Christians), but can rather be interpreted as a clue to how various natural-cultural worlds have been constructed throughout time, for better and worse. Contemporary indigenous and nonindustrialized ways of being in the world are just as valid as postindustrial globalized ways of being in the world. Traditional ecological knowledge creates different but equally valid truths as those found in modern science. Past constructions of reality provided different, but equally valid, constructions of reality or ways of being in the world. Furthermore, past ways of coconstructing meaningful worlds *reverberate* in the present and enable us to imagine together a future. These religious imaginings, very matter-real, do not become justification for a way of being in the world, but become part of the contemporary effort to imagine again how we want to become. There is no pregiven end: projections and imaginings do not close off reality, but rather hold reality open toward a radically unknown future.

I cannot stress enough the fact that this type of agnosticism is not atheism. Robust theisms of the systematic or fundamentalist stripes and robust atheisms of the Richard Dawkins type both say way too much about that which we do not know: namely, that the places beyond our sensory capacities are filled with either the foundation of God or nothing. This is precisely an illustration of the problem *of* using foundations to take us outside othe rest of the evolving, open-ended natural-cultural worlds. Neither side will ever listen to the other, rather both will try to make the other conform to her way of thinking. What other options are there from within a closed system of thinking? How is any recognition of the truth claims of an other *really* possible when meaning-making practices are seen as given, whether that means *given* in the form of nothing or ultimate meaning.

A viable agnostic theological method does not take meaning away from the world, but has implications for the scope and scale of our meaning-making projects. In foundational understandings of reality, human meaning must be narrated across the span of the universe, from origin to end. In circular understandings of time, again, meaning is all-inclusive: everything has its proper place and is accounted for. The reverb sense of time implied in this theological method is more akin to what Keller describes as fractal-like or what Deleuze and Guattari describe as rhyzomatic. In Keller's understanding of creation as plurisingular, time is fractal-like or spiral-like; it is not mere cyclical repetition nor a linear movement that cuts creation into isolated moments, but spiraling recapitulation. Just as Deleuze and Guattari's notion of rhizomatic thought works against the tendency to think in arboreal, linear terms, so Keller's understanding of time could be described as rhizomatic. In this understanding of time the scandal of particularity becomes the norm of particularity: "Any event, every space-time of the capacious process of creation, might become readable as a unique, holy, temporary embodiment of the infinite."[42] Thus the universal *is* the particular—all the particularities in the becoming creation are universal. In other words, there is no outside space. Following Derrida, Keller notes, "Nothing left outside: this might entail simple atheism and the collapse of meaning along with the space of transcendence. Or it might mean that the divine can no longer be situated in a standard theistic beyond and therefore that it cannot offer the false sense of security encoded in the classical construct of the changeless transcendence, with its omnipotence that could unilaterally intervene to reward, punish, or rescue."[43]

What I hope for in this meaning-making method is something like an interdisciplinary conversation process in which we make explicit our implicit practices of meaning making and begin to take responsibility for them. In this way we can begin to take note of those things that are left out of or harmed by our systems of meaning. This is akin to the pragmatist vision of a John Dewey or a Richard Rorty. Furthermore, it turns meaning-making processes into a participatory democracy rather than an elitist trickle-down type of activity.[44] Just as science cannot dictate nature, so here religious leaders and theologians cannot dictate meaning. Nor can any private belief dictate religious meaning. As Rorty notes, beliefs are never private: "There is no way in which the religious person can claim a right to believe as part of an overall right to privacy. For believing is inherently a public project: all us language-users are in it together."[45] Rather, both scientists and theologians help us to tussle with what these tropes of religion and nature might mean. Deleuze and Guattari also capture the type of interdisciplinary knowledge production I am hoping for in their understanding of "nomadic thinking."[46] Though we travel in time and space on textured grounds, our thoughts never land on terra firma, because the ground, as Keller notes, is precisely not foundation.[47] It is always shifting and moving. If we do not recognize this locatedness of thought, the towers of language and symbols we decide to stay in will mistake shifting grounds for foundation and then begin to remake and force all reality into the image of itself.[48] This is precisely the reason that a *viable* theology must be agnostic.

Any new symbolic constructs, including theological ones, will be located in a biohistorical context and thus will suffer from certain blind spots and the need for continuous transformation. This does not mean that we are *reduced* to the specific biohistorical context in which we find ourselves at any given moment (another form of foundational thinking), but that it is the only space from which we can see at all. There are temporal horizons between which the present exists: we can see into the past and look toward the future but never beyond to any foundational point.[49] However, we can never seal the present off from the ongoing process of continuous creation. In this sense, the discipline of theology has to be thought in terms of the changing environments around us, as a piece of environmental history rather than as somehow merely imposing thought onto matter.

In terms of a theological epistemology, this dialogical thinking involves, first, recognition of the location of rational thinking within the context of evolving epistemic communities as a human capacity rather than some acontextual space that we partake in. J. Wentzel van Huyssteen calls this rational capacity the human ability to "make sense" of and "cope" with the world.[50] Kaufman refers to this capacity, which extends even to theological reasoning, as the ability to critically converse.[51] Theological reasoning as convers(at)ion implies that there are multiple dialogue partners in the process of theological thinking and knowing. Etymologically, it also relates to the willingness to be changed (to live *with*): con-version. This openness to the other is the place from which dialogical theological exploration can begin. Rather than coming to the conversation in a monological way, with foundations that cannot be moved but must be accepted, we can come to the conversation with openness to the possibility of being changed by the other, which always already happens (whether recognized or not).[52] However, openness to the possibility that theological exploration might convert you is not the sole purpose of theological knowing. Rather, theology has always also been about making sense—making meaning and value—of the world, and this world from a human perspective is a natural-cultural world.

Given that we exist in relationship to human and earth others, this means that theological reflection and any reflection on meaning must also include our whole experience and the experience of others. One objective of this viable agnostic project is to expand theological and meaning-making practices beyond the individual, group, or humanity in general to all of creation, with the recognition that we can't get beyond epistemic anthropocentrism and that we are cocreators with all other life of the worlds in which we live.[53] Again, some type of agential realism of the type Barad offers, that understands humans in an active continuum with the rest of emerging phenomena, is the type of planetary onto-ethics that we need. "Ethicality is part of the fabric of the world; the call to respond and be responsible is part of what it is."[54]

For Christians, theology will always be about the experience *of* at least Christian humans but that does not mean that it is only *for* Christians or even humans. Just as for Buddhists, meaning will always be about the Buddhist experience, but this does not mean that it is only for Buddhists. This is precisely *not* a call for making the rest of the world conform to a given religion or system of belief, but rather a call for those working from within specific meaning-making structures to take note and responsibility for the ways in which their meaning-making practices affect the rest of humanity and the rest of the natural world. It is precisely contextual in that it argues against both: a) universalizing/colonizing the world with a particular idea, thought, or imagining; and

b) relativizing one's own way of thinking and imagining to the extent that one does not even take responsibility for how it affects others.

Again, I am extending theology as projection beyond the perhaps misinterpretation of Feuerbach's project as anthropological and somewhat individualistic version of projection. For Feuerbach, "God as such, the one universal God, from whom the bodily, sensuous attributes of the many gods have been removed, does not transcend the genus *homo*; he is only the most objectified and personified generic concept of mankind."[55] The reason for extending theological projections beyond the personal is that the worlds in which we live are much more than the personal. Our theologies are *for* the evolving nature-cultures in which we live, though they, like all systems of thought, are epistemically located in specific biohistorical human experiences. Like all other symbols, theological ones shape the ways in which our nature-cultures evolve. Though intangible in terms of efficient causality, theological constructions actively shape our worlds. Theology is projection, but projection that is not merely individualistic; it is a continuous projection of evolving communities' experiences of the world that *matters*.

Some may still argue that the method I am articulating is an atheistic one, just as some interpreted Feuerbach's claims about theology. However, again, I contend that atheism and theism are really like two sides of the same coin, as are reductionism and holism or relativism and universalism. The former of each pair reduces reality to the self, at least in some versions, while the latter makes self-knowledge knowledge of all reality. In the case of the former, we are each islands of reality, and the claims of others do not challenge the claims we make from our own islands. In the case of the latter, for instance, we are made "in the image" of God and through theological inner-course become like the omni-God we imagine, ordering all of life according to this knowledge. In both cases, we are taken out of the depths of ongoing creation, both in a way reduce all to self. As Mark Taylor has suggested, "If the master is God and the slave man, then man's murder of God is an act of self-deification. . . . The death of the sovereign God now appears to be the birth of the sovereign self."[56]

In a theology that doesn't stop at projected foundations, everything comes from something. Thoughts and knowledge have come with textures that can be reflected upon, interpreted, and reconstructed as something new, even if not ex nihilo new, for the present biohistorical context. "My" theology is not then my own projection: it comes from centuries of others both human and non, as the nonhuman other always already shapes and enters into our theological reflections. I am response-able, and in many cases responsible, in this reconstruction project, but not fully in charge or fully in the know. Because the process of knowing is always already from within specific biohistorical contexts, our knowledge systems are always interpreted anew. We cannot know how our reconstructions will affect all the life around us, but we can be open to the process of continuous creation, open to being changed by others in conversation, and open to how our theologies affect earth others.

Religions as Lines of Flight: Embracing Polydoxy

Imagination is what goes beyond the edges of certainty, which is technically the opposite of faith, in search of new ways to incorporate emergent newness into our understanding. Such imaginative meaning making is precisely not a closed system, making everything conform to predetermined categories, but rather is open to the creative, multiperspectival process of life that always exceeds our most certain categories. From an emergent perspective of newness that relies on a viable agnosticism, many possible imaginative lines of flight become better than any amount of certainty. Remember, certainty at the edges of our knowing serves to reify life in a form of conceptual violence that ends up violent toward many planetary bodies. From a substance-based metaphysics where truth transcends contexts and dictates reality, universals and orthodoxy is the name of the game in meaning-making practices. Such orthodoxy ensures the preservation of essential identities. However, if our reality is nonsubstantial, interrelated, and always in process, such orthodoxy creates the illusion of preservation of essences to the detriment of many other planetary bodies. As Keller notes, certainty has created much more violence and destruction than uncertainty.[57] Thus, from this postfoundational, emergent, and viably agnostic understanding of meaning and meaning making, contextuality, multiperspectivalism, and polydoxy are valued over universality and orthodoxy. Here I want to discuss two religious resources for dealing with multiperspectivalism and polydoxy. One comes from two philosophical tenets of Jainism, the other from a more recent movement in Christian theology.

As the reader may know, Jainism is a religion of India that emerged out of the Vedic background roughly around the same time as Buddhism. In fact, some would argue that it is not until the Buddha and Lord Mahavir begin to critique and transform Vedic cultures (roughly twenty-five hundred years ago) that you get the emergence of Buddhism, Jainism, and Hinduism. This is similar to the argument that Christianity and contemporary Judaism are mutually articulated in the backdrop of late antiquity. In other words, these religions codefine and shape one another. They are not isolated traditions. In any event, Jainism is most well known for the strict concept of ensoulment of all life and its corresponding ethical system based on *ahimsa*, which Gandhi made popular in his nonviolent resistance to imperial Brittan. Here, however, I focus on two components of Jainism that promote multiperspectivalism and polydoxy in meaning-making practices: *anekanta* ("nonabsolutism") and *syadvada* ("relative perspectives").

Anekanta is the idea that no one person can grasp absolute reality. Granted that devoted Jains would argue this applies to everyone except the enlightend *tirthas*, or teachers such as Lord Mahavir, but here I would argue it offers a deep ethico-ontological insight into our ways of knowing and making meaning out of our lives. It suggests both that no one person, much less human beings in full, can exhaust the reality of the world. No knowledge claim can grasp reality in full: there will always be something left out and not grasped. This is very similar to the differed/*différance* of deconstructionism and the abject of identity formation in queer theory. As such it leads to the idea that multiple, relative, and, I would argue, contextual perspectives must be considered when making meaning.

Syadvada, often thought of as relativity of perspectives, but here perhaps better understood as multiperspectivalism, is the idea that multiple truths can exists at any given time or place. Such multiple truths might be thought of as Deleuzian lines of flight or pathways forward into the future that might lead to different realities. Such multiperspectivalism suggests that truth is not pregiven, but that truths are cocreated as regimes of truths through the embodiment and realization of imaginative lines of flight. Again, this is not a traditional approach to *anekanta* or *syadvada*, but one that tries to bring these concepts into a discussion of a postfoundational, emergent, and viably agnostic method for meaning making. The recognition that there is not one absolute truth that can be grasped and that many perspectives on reality, both human and nonhuman, ought to be considered suggests practicing a form of polydoxy.

Polydoxy has become a new method for understanding not only the history of Christianity in particular but also religions in general over the last two decades. In conversation with postmodern theories, and in opposition to the radical orthodoxy movement, polydoxy suggests that there has never been and never will be a single orthodox interpretation of religions and/or the Scriptures and theological-philosophical traditions on which they rely.[58] Rather, claims to orthodoxy are much like power-filled claims to truth that seek to obscure their own sociohistorical constructions by claiming a transcendent foundation. Though this is not the place to go into various examples of polydoxy,[59] I will leave you with one quote on rethinking divinity, truth, and meaning in terms of multiplicity. Such work finds many points of connection with the type of negative, apophatic, agnostic, hybrid, queer, multiperspectival, embodied, and planetary meaning-making practices that I am calling for.

> Multiplicity is a dialect of porous openness, implicating a divinity that is streaming, reforming, responding, flowing, and receding, beginning . . . again. Another way to look at this is that God inter-courses . . . promiscuously. Divine promiscuity is an economy of "more than enough," but it is also a negative gesture. There is no "control" that doctrine can place on divinity, especially in the theory-resistant multiplicity of divine immanence.[60]

Throughout this chapter I have argued that religions and meaning-making practices are more like various imaginative lines of flight that become crystalized into truth regimes. Out of these regimes of truth are formed rituals and habits that instruct our performances of meaning-full practices. We as individual subjects reenact these often hybrid meaning-making practices and have some room to per/vert those practices that are oppressive toward other planetary bodies. As such, human beings are able to respond to the meanings that have shaped their own identities, which are to varying degrees responsible for cocreating future imaginative meaning-making practices that help to bring about the flourishing of the planetary community not just the human community. Our meaning-making practices, then, are central to forming our identities, and these identities are always in the process of becoming with human and earth others. I turn now to a sustained critique of identity formation and identity politics in chapter 4.

4 DESTABILIZING IDENTITY

Beyond Identity Solipsism

> Figures such as the self, the other, this nation, that god must be thought not as substances to be torn down, but rather as myths to be interrupted.
> MARY JANE RUBENSTEIN, *Strange Wonder*

If we are working from the assumption that religion and science, the ideal and material, and values and facts always already mutually inform our ways of knowing and becoming in the world, and that the realities these different concepts describe are nonsubstantial, evolving, and always open toward an unknown future, then the ways in which we think of our own selves must also be nonsubstantial. In other words, the "I" to which we refer is not some sort of essence, whether that be based upon materiality (e.g., genetics) or immateriality (e.g., the *imago dei*, cogito, or soul). Such essences reify our own identities in the face of the other and erase the very conditions that allow for an individualistic, substantial interpretation of the self in the first place. In other words, from a substantial understanding of the self, the individual backgrounds all the human and earth others that allow for the assertion of that self. The "I" turns to shadow, denies, backgrounds, and ignores all the others that go into its very construction.[1] In this chapter I begin to destabilize the myths of the liberal, Lockean "I" that has been at the heart of modern, and even postmodern, understandings of the self. Such an "I" is in many ways dependent upon foundations in the material and/or ideal realms that are separated in Western thought through Plato's myth of the cave, through the Christian monotheistic God that creates ex nihilo and the subsequent human made in the image of this God, through the scientific revolution and its mechanization of nature and location of value in the human "I," through Descartes's cogito in the Lockean liberal self that transforms dead matter to property through labor, and through the postmodern self that creates his own reality. Such separation, in other words, runs deep in the historical myths that have made up the mythical place called the West. Here, however, I argue that this separation always already ends up in an identity trilemma akin to Agrippa's trilemma already articulated. This trilemma forces the "I" into identity solipsism, or so I argue. In order to destabilize this "I" so that it might break out of this trilemma, I will offer specific examples of identity monism, identity idealism, and identity materialism to show how they all end up assuming a substantial material and/or ideal world. This will set the stage for a fourth option of identity formation, which I call emergent identities, in chapter 5. What I suggest is similar to what Mary Jane Rubenstein suggests in her book *Strange Wonder*: that at the base of all of our identity constructions one does not find a certain, substantial foundation, but the "gaping wound" of wonder.[2]

THE IDENTITY TRILEMMA: IDENTITY SOLIPSISM AND THE COLONIAL "I"

The identity trilemma I describe here is similar to that of Agrippa's trilmma. However, I will identify three problematic categories of the identity trilemma—monism, idealism, and materialism—and opt in chapter 5 for emergent identities akin to the solution of infinite regress discussed earlier. All three of the former options end up in solipsism. In other words, identity monism, idealism, and materialism depend on some type of foundationalism or circularity. Foundationalism and circularity are two sides of the same coin, which end up in solipsism: that is, they both end up creating the world in the image of their own "I." As such, identity monism, idealism, and materialism end up in the same place: as a monological "I," creating the world in its own image. The way out, a type of infinite regress, will be offered in the next chapter. Here I lay out the terms of the identity trilemma, providing specific examples along the way. Let us turn first to identity monism.

Identity Monism and the Gay Marriage Debate

> On the one hand, from a queer perspective, we learn that the dominant culture charges queers with transgressing a natural order, which in turn implies that

nature is valued and must be obeyed. On the other hand, from an ecofeminist perspective, we learn that western culture has constructed nature as a force that must be dominated if culture is to prevail.
GRETA GAARD, *"Toward a Queer Ecofeminism"*

Starting with the position of the trilemma that forces the "I" to be schizophrenic through backgrounding change altogether, as the epigraph to this section intimates, identity monism is the perspective that places foundations in both the material/natural and the ideal/cultural realms. In other words, in the identity monist perspective there is universal right/wrong, good/bad, truth/falsity, and these categories are unchanging for all times and places. In addition, this foundationalism in the ideal, cultural, religious, meaning-making world is supported by substantial foundations in the natural, material, or physical world. What is right is both God-given and natural. What is wrong is against God and unnatural. Put in another context, what is true is available to all through reason and found in natural laws in the material world. What is untrue is unreasonable and against the laws of nature. From the monist perspective, then, the meaning-making realm and the natural realm converge on the same reality.[3] Finally, categories of race, sex, gender, and sexuality are all foundational and support one's understanding of the "I" as being both right and natural. Other versions or transgressions of these categories in relation to the monistic "I" are thus wrong and unnatural.

Identity monism is then solipsist in that the "I" is never challenged by any other. There is no dialogue with the outside world, but only the categorization of the outside world into these strict categories that conform to the "I." This is why I argue along the same lines as Greta Gaard in the opening of this section that identity monism ends up in schizophrenia. In order to justify its own identity categories, the identity monist must code nature as both good and bad, all the while covering over these shifts as if they were completely consistent in making claims about nature and culture. When a heterosexist and patriarchal approach to identity monism is imposed upon the world, for instance, queer identities are unnatural/bad, thereby suggesting that nature is inherently good, while female identities are closer to nature and in need of subordination to men, thereby suggesting that nature is inherently bad and in need of the light of culture or reason. Or, we might think of the traditional understanding of patriarchal culture in the West in the nineteenth century. At one and the same time the masculine space is identified with public/culture and the feminine space is identified with personal/home, while the public/

culture space is also charged with turning men into sissies or dandies, hence the need for boys to go into nature to reconnect with their true manhood. "White men came to assert their increasingly heterosexual identities in the wilderness explicitly against the urban specter of the queer, the immigrant, and the communist."[4] In both cases I have mentioned, the decision of the "I" is seemingly consistent from the "I"'s perspective, while a schizophrenic decision about good/bad, right/wrong, natural/unnatural is imposed upon the other. Such is the case today with identity politics on both sides of the gay marriage debate in the United States.

On the one hand, you have the opposition to gay marriage that argues that gay marriage is both unnatural and ungodly. Eisegetically quoting about eight different sources in biblical Scripture, some self-proclaimed Christians argue that it is an abomination according to the Bible, which is then understood as the word of God. Note, it is not only Christians that quote Scripture in this literal way, but I am focusing here on the context of the United States and the type of foundational thinking most prevalent in the U.S. is some version of Christian foundationalism.[5] Other cultures and religious histories have different ways of constructing sexuality from that of the West: gender dimorphism is not inherent to the animal kingdom (including humans), but is rather, often imposed on the world by Western colonizers.[6] So-called third genders and same-sex love may be seen as a part of one's life in various cultures, but not as the defining point of identification.

I am also not suggesting that all Christians, much less all religious thinking, use Scriptures or religious thought in this foundational way. For those that do use Scripture in this way, there is often no amount of exegetical work that can be done to convince them that the bible is always already interpreted and that those interpretations change over time. Not to mention that something like homosexuality didn't even exist during the time in which biblical Scriptures were written down and thus there is no Hebrew or Greek term for homosexuality. The point is that from this foundational perspective if God says X, then X is right or wrong (in this case, wrong). At the same time, these same people will say that gay marriage is not right because it is not natural. At times the naturalness of heterosexuality is argued along religious lines (e.g., "God created Adam and Eve") and at others along the lines of biological procreation: obviously, if sex cannot result in offspring it must be unnatural. Justified, then, by both God and nature, the solipsist conclusion that gay marriage is wrong cannot be challenged by any other.

On the other hand, you have proponents of gay marriage following this same logic. Many LGBTQ peoples find themselves struggling with how to navigate between the often negative ways in which their religious traditions have treated sexualities other than heterosexuality and their own newfound homosexuality or bisexuality. This struggle is not to be belittled, and I do not challenge the ways in which any one person heals/reconciles her sexual identity with her religion. It is a struggle and challenge that I commend, for it is all too easy to choose *between* religion and coming out. In any event, there is a certain rhetoric behind those that are pro-LGBTQ marriage that suggests "God made me this way" and "multiple sexualities are natural." Thus one's sexuality is both natural and God given. Seemingly contradictory to the way in which *Evolution's Rainbow* queers nature, attempts to discover or found human homosexualities and bisexualities in nature are just as foundational as those that attempt to found heterosexuality in nature. Likewise, the queer theory and LGBTQ exegesis that has taken place in religious studies over the past thirty years or so can easily become foundational when the process of interpretation is covered over with statements such as "god did" or "god said." Equally foundational and opposite their opponents, identity monists on the pro-gay marriage side will not be challenged by any other.

The problem, from my perspective, is that such identity monism covers over the structural—historical, political, economic, and legal—systems in place that background uncertainty in an effort to equate one way of being in the world with *the* way. On the part of anti–gay marriage monists, there is no recognition of the changes in what marriage has meant throughout the centuries. It was basically an economic arrangement whereby a woman was given to a man. Most opponents of gay marriage, though not all, do not acknowledge these historical roots of marriage, nor do they acknowledge the different ways in which marriage has been conceived throughout history and in different places, because this would admit to the construction of marriage rather than its naturalness or God givenness. Similarly, in one's interpretation of a form of sexuality as natural he must ignore a whole host of sexualities that exist throughout the natural world. On the pro–gay marriage side, there is little to no recognition that what is being argued for is the right *to be like* heterosexuals. In other words, why do all loving relationships necessarily have to end up in the form of marriage? Might it be that this form of pairing has been historically, politically, economically, and legally privileged? From this perspective, instead of "can the subaltern speak?" we might ask, "can those in a GLBT relationship speak?"[7]

I would argue that inside every LGBTS (lesbian, gay, bi, trans, straight) there is a Q (queer) waiting to get out. We are all way too complex to fit into any identity fully. There is always a remainder or leftover part when norms are applied to bodies. Yet from an identity monist perspective there is no possible way to allow for expressions of something other than what is historically, politically, economically, and legally privileged. These are the systems that set up the "I," and they cast a shadow over their own creations, thereby making them seem natural and God given. As Rubenstein notes, "Like any good master, consciousness predicates its integrity upon the enslavement of all that might threaten it."[8] Rather than challenge the heteronormative capitalist politics of love (more on this later in this chapter) and argue for free health care, free education, changes in inheritance laws and legal guardianship of children, it is easier to just conform to heteronormative relationships. This identity politics–based struggle ends up allowing some people into an already broken system rather than challenging that system altogether.

Furthermore, this identity-based political struggle does little to challenge the idea that heteronormative monogamy is the only legitimate type of relationship. This heteronormative monogamy as the norm keeps us locked into the capitalist politics of love as possession and, as I argue in chapter 6, helps to create a monogamous relationship to place that may actually exacerbate ecological and social ills in a world that is marked by movement and change. Though I will discuss this further, let me offer up an example of what I mean here.

Our legal system and our religions tend to favor monogamy for historically economic and procreative reasons. Such favoring of monogamy has led to a situation of straight supremacy equal to that of class, race, and sex supremacy.[9] As Dean Spade argues, we have a "legal system that was formed by and exists to perpetuate capitalism, white supremacy, settler colonialism, and heteropatriarchy."[10] Monogamy, just like celibacy for priests, is wrapped up in power-filled institutions. Monogamous marriage was historically based on men owning women and ensured that inheritance would pass from biological father to biological son. In other words, it was an economic institution. Far from being founded in the teachings of any religion or Scripture, marriage as a practice was first a legal institution and was adopted by the early church in Western culture. Family values, as so many have argued, are nothing but Victorian values wrapped in religious garb.[11] The historical ideal in many religions—including Christianity, Buddhism, and Jainism—is actually celibacy, not monogamy. More specifically, in Christianity the teaching of Christ is for his disciples to give up mere blood kinship for a broader understanding of family. Such admonitions stem, perhaps, from the realization that monogamy and bloodline-based notions of family may actually lead to localization and narrowing of moral concern: I am concerned with what is mine and about my own rather than some of the larger problems in the world. This does not break free of the logic of ownership and possession, even if we shroud such an ideal in the garb of love and equality. A true signal of such possession is the assumed normalcy associated with feelings of jealousy. As the authors of *The Ethical Slut* suggest, "Many people believe that sexual territoriality is a natural part of individual and social evolution. If you believe that, it's easy to use jealousy as a justification to go berserk and stop being a sane, responsible, and ethical human being. Threatened with feeling jealous, we allow our brains to turn to static with the excuse that we are acting on instinct."[12]

In other words, jealousy is justified as a natural response when monogamous marriage is God given or somehow hardwired into our identities. I would argue that such jealousy reveals more about the lingering notion of ownership in romantic love. In other words, reflect on jealous feelings you may have had toward a lover, and now reflect on something being stolen from you or a trespasser on your property who does not respect it or someone who gets the job that you almost felt certain was yours, and I would wager that those feelings are all very similar. More on this will be discussed in terms of the monogamy of place in chapter 6, but here I want to offer an alternative.

First, and before I am charged with being antimonogamy, I have to say that monogamy is not an inherently bad thing, nor is marriage or heterosexuality for that matter. It works well for many people. The problem is that it should not be enforced on all peoples and relationships through religious, legal, economic, and social structures that favor monogamous marriage over other forms of relating. Again, the *institution* of heterosexual monogamous marriage—not to be confused with the individual choice or desire for heterosexual monogamous marriage—works well in the contexts out of which it emerges: historically rural, relatively isolated populations that need large families and rules for maintaining kinship relations and societal relations. Fast forward to the contemporary context: predominantly urban populations, moving around the globe through increased speed in communication and transportation technologies, and people that have little need for large families or even procreative sex due to the advancements in technologies of reproduction. Simply put, we get the chance to meet a lot more fish in the sea these days than our predecessors ever did. We live increasingly in a planetary community and our identities, far from being stable, as identity monism would suggest, change with contexts and with relationships. Even neuroscience tells us that our neurons are pruned by every interaction we have with human and earth others. What I am suggesting is that the "you" in San Francisco may not be the same "you" in Miami or Berlin or Yogyakarta, Indonesia. The contexts you are in—including humans, other animals, and the bioregion or ecosystem—interact with you to coconstruct your identity. In such a fluid world, a belief in or adherence to concepts of stable identities, whether monistic, idealistic, or materialistic, ends up backgrounding the various differences of self and other in contexts other than one's idealized place.

Our monogamous relationship to place, identity, and one other (till death) end up making negative abjections of the various others that codefine who we might become in any given context.[13] We literally reject full engagement with others and, in the process, repress the multiple selves that we might become and project those parts of ourselves onto the others that are abjected. What I would suggest, then, is that we need religious, legal, and economic institutions that promote polyamorous relationships. An affirmation of polyamory takes love out of the zero-sum game of consumption, fear, and limitation and turns it toward more of an art form, such as is found in the Kama Sutra. Love, far from being a scarce resource, is one of the few renewable resources around and one that can actually cut across many diverse boundaries such as self/other, human/more than human, and old/young. Love, rather than being understood as a dangerous threat by institutions of hierarchy and control, might better be understood as a subversive art form.

"Understanding love as an art helps to see longevity as a virtue, for as one learns more one becomes a better teacher. It also helps young people take good care of their well-being as they look forward to a beautiful and joyous future when they will be older and better artists of love too. Love is free, and it can be multiplied at will. It is a renewable resource that saves one from the trappings of useless consumerism."[14]

This is not to suggest that everyone needs to be sexually polyamorous, but we need institutions that support polyamorous affirmations of planetary others. Such polyamory allows us to practice "other love" that is so prominent in the world's religious traditions. We can begin also to understand that our very own identities are coconstructed by earth others and are constantly shifting and evolving over time. This polyamorous affirmation of planetary others affirms our rootedness in evolutionary and cosmic structures, as well as our rootedness in contemporary human histories and structures. Heteronormative, monogamous marriage, then, simply does not capture the multiple forms of possible relationships, loves, and interactions in the planetary community. Monogamy, serial monogamy, open relationships, celibacy, multiple loves, one-night stands, sibling friendships, communal living, orgies, and companion species, just to name a few, might all be legitimate forms of relating to others in various contexts. Of course, such relationships will look differently depending on whether or not loving relationships involve children, pets, and other dependents. I am not suggesting here that responsibility for dependents be shrugged by any means, but such multiple pairings would surely require a wider sharing of responsibility for dependents. Such a sharing of responsibility moves beyond trends in education and laws that often give parents ultimate rights over the content of their children's education, how they are disciplined, and what belief systems they are taught. Such narrowing of responsibility for future generations to parental prerogative denies the social and ecological embeddedness of future planetary citizens. This narrow view of responsibility for future generations has economic roots in the Lockean liberal self.

When, through identity politics of various types, we narrow ethical consideration down to individual human rights (not that these are unimportant), we play into the Lockean liberal understanding of the individual and support monistic understandings of identity without challenging the larger institutions that keep us locked in to these ways of thinking about the self. Individualism of this type is politically, legally, and economically expedient. Yet it comes at the cost of backgrounding the planetary others that constitute any given identity. As Val Plumwood notes, "Humanism must come to terms with an affirmation of the denied nonhuman side of the dominant human culture that is labeled nature if it is ever to find a satisfactory form for its human application."[15] Identity monism is supported through foundational thinking about both thought/meaning/religion (the soul or essence of the self) and nature/the material world (it's in the genes, hormones, neurons). In various ways, each of these foundations has been critiqued, but if both foundations are not challenged, identity solipsism still results. It is to a discussion of identity idealism and materialism that I now turn.

Identity Idealism: Mind Over Matter, the Intersexed, and Third Genders

Identity idealism is the type of identity solipsism that occurs when one considers the material world as dynamic, chaotic, malleable, and ever changing and the ideal world as ordering, static, or foundational in some way. Included in this type of idealism would be Cartesian understanding of the self—"I think therefore I am"—and the Lockean liberal self. Remember the Lockean self is the *tabula rasa* or blank slate (body/brain) that can be shaped and molded by culture, just as the individual creates value (in private property) by mixing his labor with dead matter. Also included in this type of identity idealism would be the postmodern constructivism that understands the natural world to be completely constructed by human language, reason, culture, and meaning. If the idealists only find foundations in the meaning making, religious, cultural, ideal side of the ideal-material equation, then why do they also end up in solipsism?

The problem is that a foundation in either the ideal or the material realm will act as an anchor for meaning that will then distort all others flowing past the anchored identity. In other words, as Rubenstein argues, following Jean-Luc Nancy, any essential identity distorts reality to confirm its own self-justified identity. "'Man,', this race, that nation, this idea, that teleology—all these mythic substances effectively wage war against *existence itself* in order to posit themselves as infinitely self-(that is, un-) relational."[16] This anchored ideal is always already anchored in relationship to materials. There is no ideal without material nor material without ideal; thus bodies are cut, strangled, and made to conform to the contours of the anchored "I" as they flow past it. A great example of this identity idealism is what happens to intersexed peoples and the "third genders" of many cultures when gender dimorphism is imposed upon real bodies.

At the risk of sounding like I am trying to found the idea of multiple sexualities in nature, the rate of humans born intersexed is around one in one thousand. Historically, in places that conform to some type of gender dimporhism, these children have been operated on at birth to conform to one or the other sex: male or female. There is a lot of pressure on parents to choose the sex of their child, as if this had absolutely nothing to do with the unique biological assemblage that made up the child.[17] Such pressure belies an inability to imagine life outside the gender/sex binary. That material which does not conform to this binary must simply be molded to fit. Needless to say, these post-op intersexed children forced into sexual submission in the image of male or female sometimes end up having all types of problems as they grow older and into their evolving sexual selves. In other words, despite the construction of childhood (in the West especially) that removes children from labor and sex, their sexuality and labor practices are imposed upon them by making surgical decisions at birth.[18]

Currently, there is a growing group of intersexed individuals and medical professionals that argues for intersexed rights: that is, to not perform operations until the person in question chooses whether or not s/he wants to be operated on at all. It goes beyond allowing the intersexed person to choose either male or female by allowing the intersexed person to choose whether or not to conform to a sex at all. Cisgendered people, or those whose gender identity and sex identity match, may not be able to fathom such transgendered realities, but that is precisely because cisgendered realities are supported by legal, economic, social, and religious institutions, just as is sexism or heterosexism.

To point out the schizophrenia of this gender dimorphism founded in identity idealism, one just needs to consider the reaction of heteronormative, patriarchal society when a child born intersexed is operated on versus when a grown adult decides that she or he wants to change her or his sex. Both are cases of being transgendered in some way or being born a sex other than the gender you identify with. In the first instance, the surgery is required to make the person right and in the second instance, the surgery is often looked down upon as wrong. On the other hand, in some cases where third genders are accepted, such shifts are accepted precisely because one is understood to transform into female or male (more on that further on in this chapter).

I should note here that one could also argue that someone seeking a sex change is caught up in identity idealism: imposing an ideal upon a body that does not match up. This is, indeed, a form of identity idealism. Note that I am not arguing that we can escape the ideal side of identity construction. We are all caught up, to some extent, in identity ideals when we try to make our bodies conform to an image through working out, makeup, wearing certain clothes, piercings, dieting, and getting tattoos. The problem comes when one argues that everything *must* conform to this ideal: in this case, gender and sex dimorphism. This is often not the case with those in the trans community who often identify as trans and exist along a spectrum of bodily modifications from cross-dressing to hormone therapy to partial and full operations. It is also sometimes the case that people who once identified as trans and then fully transition consider themselves to be a full member of the sex/gender they transitioned to (and no longer as trans) after going through a transformational or transition period. In other words, the transformation is chosen and recognized to a certain extent. In the case of the intersexed who are operated on at birth, no transition is acknowledged and the gender is imposed from birth, covering over any "abnormalities." In other words, their trans identity is written over as cisgender, and they are left to pick up the missing pieces as they age and begin questioning both gender dimorphism and their own cisgender status.

Yet another example of identity idealism is the inability to see anything outside of gender and sexual dimorphism. Many South Asian and Southeast Asian cultures, as well as indigenous cultures, have what are commonly referred to as "third genders." Variously named *hijra*, *waria*, *nadle*, and two spirits, among many others, these identities are often so far beyond an idealized gender and sexual dimorphism that many associate these identities with the LGBT community of the Western world. Again, such associations are a form of colonization, or the imposition of sameness upon all other locations; here the imposition of anything other than heteronormativity, cisgender status, and gender dimorphism as LGBTQ regardless of histories, traditions, and cultures.

One such example is found among the *waria* of Indonesia. Many Western anthropologists studying indigenous, Asian, and other cultures around the world, and many that identify with the Western categories of LGBTQ, are tempted to read such queer identities into other cultures and histories. On the one hand, this reflects the elision of the construction of sexuality as an identity in the West.[19] Such a construction of sexuality imposes the idea of internalized sexual identity upon all peoples around the globe who do not fit into the conception of heteronormativity. As Tom Boellstorff notes, "This ontology of the closet draws heavily from a Christian metaphysics construing the transient body as secondary to the everlasting soul."[20] Rather than such an identity, Boellstorff suggests that in places such as Indonesia, and I suspect in many other places around the world, multiple selves without integration can be experienced without a problem. He describes such self-formation as the archipelagic self. The archipelagic self is "not predicated on a singular selfhood that coheres across time and space, but" is "capable of movement through different 'islands' of life that do not need to resolve into one."[21] In other words, there is not the drive among gay and lesbi Indonesians to come out of the closet or to have a single identity based upon their sexuality. "For gay and lesbi Indonesians, the self is not that which moves from island to island; it is the water itself, lapping up on multiple shores at the same time."[22] Such an understanding of multiple selves matches for the most part my own experiences in Indonesia. Many of the lesbi/gay peoples I have met there over the years intend to have a heterosexual marriage with children at some point. Furthermore, most of them do not feel the need to come out to extended family (who often live in rural areas), but rather maintain different identities. Granted, many Indonesians are now also identifying with the Western

LGBTQ movement and adopting a coming out process, but the point here is that such alphabet soup ought not be imposed on the gays/lesbis of Indonesia or any other place.

The *waria*, or traditional third gender (though this term is even problematic), are yet another example of breaking out of the binaries imposed upon sexuality and gender in other parts of the world outside the West. These are men who dress and take on roles of women, but who do not identify as gay/lesbi, or even trans. They identify as *waria*. Such identities are not to be confused with the traditional *bissu/warok*, who have a long history in Indonesia as entertainers, but were first spoken of in Indonesia in large trading ports in the 1800s, in other words, in a colonial era.[23] Such hybrid identities challenges the desire to give priority to indigenous over postcolonial and global positionalities: "Chronological priority does not necessarily mean ontological priority: a subjectivity shaped by 'global' forces (like Islam) may be experienced as more foundational than one shaped by 'local' forces."[24] In other words, hybrid positionalities such as the *waria* challenge and trouble any idealized origin in time or chronology of time (more on origins will be said below).

Waria also trouble distinctions made about gender dimorphism and heteronormativity. They are not, as Boellstorff notes, considered on a continuum between male and female, but just as *waria*. They are not considered trans and in fact pride themselves on having male parts while performing as women. Men who have sex with them are not thought of as queer or having same-sex sex, but as having sex with *waria*. Such positionalities trouble the idealist identity construction that would impose an idea about identity—e.g., that gender and sexual dimorphism can shape and explain all identities, including animals and plants, by the way[25]—regardless of the differences within and between biohistorical/cultural expressions of said identities. Again, this is an imposition of an idea(l) upon bodies in a way that doesn't allow for otherness to come through, emerge, or be expressed. Such identity idealism is then solipsism. The rest of the world is made to conform to the categories of one's own identity assumptions. There is still yet a third way in which such solipsism occurs, namely, identity materialism.

Identity Materialism: The Case of Caster Semenya and the Myth of Origins

As Catriona Mortimer-Sandilands and Bruce Erickson note, "there is an ongoing relationship between sex and nature that exists institutionally, discursively, scientifically, spatially, politically, poetically, and ethically."[26] This is true of all other foundational identity markers as well, including race, gender, and sexuality. The solipsism that takes place in the case of identity materialism happens through fixing certain identity markers and categories as natural, found in the genes or neurons, or just in nature in some sort of vague way. In other words, from the identity materialist position the category of nature or material is static, leveling, and the same for all. Platitudes such as "deep down we are all the same" usually belie such identity materialism, as do attempts to found identities and characteristics of identities in the genes or in evolutionary history. From the identity materialist perspective, there is no universal culture, ideal, or value; rather there are many cultures and one nature. Much of Western modern science has assumed a similar sort of thing: while cultures may be various and manifold and diverse, there is one correct interpretation of nature, and all nature follows the same set of natural laws.[27] Sociobiologists and evolutionary psychologists are perfect examples of this type of identity materialism.

The solipsism here occurs because there is no recognition of the evolution of ideas about nature or how religion and science are always already involved in understanding the world. As such, a current perspective on nature is taken as natural and then enforced regardless of context, all in the name of development, enlightenment, or progress. Where there is recognition of subjectivity in observations, such subjective injections are seen as closer to or further from the truth of the objective material world. It just so happens that the modern perspective of the rest of the natural world seems to continually justify its own view in the face of other, more traditional views. As Boellstorff notes, "Tradition is the shadow modernity casts back in time to see itself whole."[28] In other words, the materialist perspective understands its materiality as somehow the only possible way of actually knowing the world in the face of many other lesser or outdated ways of knowing (aka cultures). For the materialist expert, all others should listen to the "I" who claims to have knowledge of the material foundation of all identities. One thinks here of the pejorative way in which Western secularism and its handmaiden modern science are often paraded about as the evident truth in the face of all other cultures, traditions, and beliefs.[29] There are, in particular, two very recent examples of the type of solipsism that occurs from a materialist identity perspective: that of Caster Semenya and the relatively recent discovery of *Ardipithicus*.

Caster Semenya is a South African middle-distance runner who won the World Championship Gold in 2009 and silver medals in the 2012 Olympics. She was consistently one of the fastest runners and had all the makings of an

Olympic champion. At least, that is, until questions concerning the genuine nature of her sex began to surface. Statements about her being a masculinized woman or an XXY, androgynous, or hermaphroditic person instead of a "real woman" flooded the international press.[30] In the face of this, she maintained her sexual nature as a female, as did her family, as did the community, and the president of athletics of South Africa at that time, Leonard Cheune. However, these verifications were not enough for the Olympic committee. In order to continue to compete in the Olympics, Semenya had to submit to verification of her sex by an international panel of Western, modern medical doctors. In response to this demand, Leonard Cheune stated, "Who are white people to question the make-up of an African girl? I say this is racism, pure and simple."[31] The charge of racism made against this international panel should not be taken lightly, especially in the context of a South African making this claim against the international community. The recent history of apartheid aside, one should remember the longer colonial history and the case of Sara Baartman (aka the Hottentot Venus), also from the region now known as South Africa. Indeed, many claimed that Semenya was and is a contemporary Hottentot Venus. Such racism may no longer be overt, but it does hide behind the foundations of what is natural and unnatural, just as all other isms have in the past. Furthermore, as we have seen, there are and have been intersexed, third and fourth genders, and two spirits throughout the histories of most cultures. At times such identities are transgendered and at other times cisgendered; that is, an intersexed biology might match a gender outside the imposed gender dimorphism. The point here is that there are many possible subjectivities that any one embodiment can take. Identity materialism refuses to realize this and requires that everyone conform to a cisgendered reality. As Boellstorff notes, we ought to distinguish between "subject positions" and "subjectivities," where "subject positions" are seen "as socially recognized categories of selfhood and subjectivities how individuals inhabit a subject position, but in ways that always exceed and transform its logic, even while being powerfully shaped by that logic."[32] This creates a proliferation of identities beyond being strapped to cisgenderism. Identity materialism, like the other types of identity solipsism, can also be well meaning and not quite as sinister as in the case of Caster Semenya, who, by the way, was deemed to be woman enough by the Western scientists. Anthropologists' and primatologists' reactions to the discovery of Ardi are a more benign example of identity materialism.

In 2009 paleontologists announced the discovery of *Ardipithecus ramidus*, aka Ardi, whose bones were first studied in 1994. Ardi was more recent than Pan-Homo LCA and older than *Australopithecus*, thereby becoming a missing link in the chain of evolution that distinguished chimpanzee and *Homo sapiens*. The partial remains of Ardi include most of the skull and teeth, hands, pelvis, and feet.[33] From these few remains, paleontologists, primatologists, and anthropologists began to suggest all sorts of things about the nature of Ardi and thus the nature of the ancestor to *Homo sapiens*. Anthropologist Owen Lovejoy, for instance, claimed that "far from being a recent evolutionary innovation, as many people assume" the monogamous pair bond "goes back all the way to near the beginning of our lineage some six million years ago."[34] As this monogamous pair bond was based upon a heterosexual, reproductive monogamous pair bond, heteronormativity is here literally founded in the natural world. In other words, heteronormativity and even signs of the Victorian era nuclear family are here located in the material fossil record. Such constructions of family, according to thinkers as varied as Michel Foucault, Karl Marx and Friedrich Engels, and Rosemary Radford Ruether, are structures that are created over time through processes of urbanization, land enclosures, the Protestant Reformation, and the industrial revolution. In other words this structure of a monogamous pair bond is created by and in turn supports a capitalist-style economy that is now being enforced all over the world through the process of globalization. Are we to now understand capitalism as having a material base in reality?[35] Here we see the struggle for territory between historians/theorists and materialists over authority. A materialist would assume something like history and theory are only necessary until a material explanation is found. However, a theorist and historian would point out that scientists always approach the material world with subjectivities that are shaped by theories, ideas, traditions, and histories of nature-cultures. Such solipsism then cuts off the critical investigation that our complex ideal-material worlds require.

Yet another interpretation of Ardi comes from the famous primatologist Frans de Waal. He wrote, "The once-popular killer ape theory is crumbling under its own lack of evidence, with 'Ardi' putting the last nail in its coffin. On the other side of the equation, the one concerning our prosocial tendencies, the move has been towards increasing evidence for humans as cooperative and empathetic."[36] Granted, pro-social and peaceful origins sound better than violent ones, but the attempt is the same: make foundational claims about identity based upon the obvious evidence in the material world. Furthermore, founding ideas about human nature in the material world are just as likely to lead to claims that we are inherently violent, racist, and sexist as we are peaceful and egalitarian. There are many other possible examples, including the search for the gay gene and the search for other natural orientations toward being smart, musically inclined, and artistic, for instance. My point here is not that nature doesn't matter, but that nature can no more be a foundation for stable identity than the ideal world can be. Such identity stabilizations fall into the trap of substantial identity formation that is at the heart of the modern liberal self, founded as it is in the split between matter/ideal, nature/culture, and in the notion of substance-based thinking in both science and religion.

This false trilemma offers only fake distinctions in identity choices. In other words, identity monism, identity idealism, and identity materialism—all three end up in solipsism: an acontextual understanding of identity. Furthermore, they end up committing what some have called the naturalistic fallacy, that is, moving from the is to the ought. One would think here this would only apply to identity materialism. Identity idealism, we might suggest, participates in the supernaturalistic fallacy by moving from the ought to the is. However, if we follow the idea that religion and science are always already together, there is no way to separate out the naturalistic from the supernaturalistic or the ideal from the material. The point is that one stakes a claim in one or the other side and then forces reality to conform to that logic. Such thinking can lead to a commodification of identity formation.

As mentioned before, substance-based identity constructions, whether monist, idealist, or materialist, all fall victim to the "ontology of the closet." In the case of identity idealism, this essence of the self is often the soul or the cogito; in the case of identity materialism, the ontology of the self is based on a material essence that resists all changing contexts. Such ontological understandings of the self abject all relations and interactions that constitute the self and thus can lead to treating others as commodities. Not only other others are made abject but also the others of the self as well: if my "I" consists of an isolated center, I will do all I can to maintain that center and resist the challenges that lead to a fluxing "I." I will spend tons on plastic surgery to resist the ravages of age; I will spend money to isolate myself from the changing environment so that the "I" can live in relative equilibrium behind closed doors and inside gated communities; I will buy certain clothes to distinguish myself from those around me whom I might find less savory; I will resist the idea that I have changed and learned throughout time by suggesting "I have always been this way." This ontological understanding of the self comes at great monetary, energy, and material costs to others. This reifies an identity and categorizes all other identities according to its own reified logic. Such reification serves an economic function. It is this economic function that I call the capitalist politics of identity construction.

The Capitalist Politics of Identity Construction

> Rather than wedding ourselves to capitalism, to the privatization of welfare, health care, and so forth through a movement for gay marriage, might it not make sense for lesbians and gays to abandon a narrowly identitarian framework and see it in "our" interests to join battles to shift the balance of power against capital?
> MIRANDA JOSEPH, "Family Affairs"

An identity based on some sort of origin or foundation seeks to get to that origin or foundation so that it might express itself in a *genuine* way. This genuine way, in order to be genuine, entails some type of singularity and some type of teleology. This economy of values then deals in authentic exchange of genuine substance. This economy of genuine exchange creates much chaos. "Indeed, the [supposed] stability of modernity has always been exceptional. The emergency and chaos that are its inevitable byproduct are the norm."[37] In other words, the development of complex systems of genuine identity that understand themselves as isolated from human and earth others are actually, as the theory of emergence and nonequilibrium dynamics suggest, made possible only through the use/entropy of many other sources of material/energy. This can also be seen in the development of a divine economy of salvation in the Christian tradition and its subsequent imperial impositions on multiple human and earth others.[38] Such energy-material-intensive stabilizations of identities and meanings then warrant some destabilization. In between the end of this chapter destabilizing identity and the next, which attempts to construct not authentic but open, evolving, ethical ecoreligious identities, I end with a suggestion for an economy of transition.[39] An economy of transition challenges the capitalist politics of love found in modern, Western, family values.

Just as we understand the work that took place to internalize women's work to the home and externalize men's work to the public sphere throughout the scientific revolution,[40] industrialization, and into the Victorian era in the way that Marx and Engels do, i.e., as created by and supportive of capitalism, we should also analyze the globalization of such family values.[41] Family values are really just values emanating from a privileged upper middle class. As many have pointed out, the nuclear family is rarely attained, even in the West. It is rather an ideal that is based upon the modern, liberal, patriarchal self. In other words, it is an externalization of the ideal, isolated individual: the male head, the female heart, and the offspring to ensure that the body (property/capital) continues to grow. The unit of concern and consumption becomes the isolated nuclear family, whose isolation makes abject all human and earth others through the resources that reify the family's isolation. In order to deconstruct these family values, rather than follow the familiar lines of Foucault, Engels, Ruether and others, I will actually suggest that the Christian right has one thing correct about family values, and that is the assumed ontology of the liberal subject.

The idea of ensoulment, as mentioned before, reifies the self and suggests also that there is a genuine, original self. Just as there is an origin to creation found in a monotheistic God that creates ex nihilo, so there is a genuine ensouled self that creates his world as if ex nihilo.[42] Such a self is fascinated not just with roots, origins, and genuine understandings, but with authenticity and originality in the things he consumes.

One example of this is found in the copyright laws that exist in Western countries: they are here to protect the original regardless of how many ideas, histories, and resources and how much labor went into that original and actually undermine its originality. A patent on a drug will earn drug companies billions, even though hundreds of years of previous research, countless plants, animals, minerals, and energy resources and millions of taxpayer dollars go into the production of this unique (not generic) drug. In other words, the relations are backgrounded in order to make out of a product an original. The same can be said for orthodoxy in religion: what religious tradition is not a hybrid of every tradition that it comes into contact with? Easter, for instance, is nothing without Passover, without German traditions of the bunny rabbit, and without Roman traditions that celebrate the spring equinox. Yet Christians claim Jesus's death and resurrection as the real or genuine reason for the season. The isolated individual seeks original things in isolation from their backgrounds.[43] Such isolated commodification/reification of identities, families, traditions, and products is precisely what capitalism depends on. These theological and metaphysical traditions help cocreate the world and universe as a Newtonian collection of objects. In other words, these substance-based ways of thinking help to create the very reified, isolated worlds of late technocapitalism.

Take, as a sharp contrast to Western copyrights and identities, the understandings of identities in South and Southeast Asia. Families and communities are hardly isolated from one another; homes are not even cut off from the outside world—often allowing many animals, insects, and other elements to flow freely through them; copyright laws are nonexistent, and a copy is just as good as any sort of genuine product or original. Understandings of the self also seem to be hybrid as Buddhism, Hinduism, Jainism, and Islam flow into one another in India or in Indonesia where Muslims may go to Hindu fertility temples and use traditional medicines from Javanese or Balinese traditions (depending on the context). Authenticity and originality are not highly sought after, and the self is much more fluid and changing according to contexts, as the notion of the archipelagic self discussed earlier suggests.[44] Family structure and relationships between family and society and self and other are thus constructed quite differently. Accordingly, the concept of the commons is also stronger in these societies—waters, lands, and other such resources often belong to the people. Nonhuman animals are also seen much more as companion species rather than mere pets or economic animals.[45] These companion species are part of the human community and are not shut out of households, work, and daily life. I suggest that such hybrid realities, which recognize the coconstructions of identities and the evolution of identities over time in different contexts, probably have theological support in the more polytheistic background of South and Southeast Asia. Of course, I do not want to oversimplify: Buddhism is more atheistic than polytheistic, and many still argue as to whether or not Hinduism is monotheistic or polytheistic. Furthermore, Islam is one of the major monotheistic faiths—yet in Indonesia it takes on a hybrid character that mingles with local traditions and past waves of Hinduism and Buddhism just as Catholicism becomes hybrid in many African, Afro-Caribbean, and Latin American contexts. Even such claims of hybridity are often made at the expense of the hybridity in the very formation of most religious traditions. In other words, there is no original to compare to the hybrid versions of traditions, but rather multiple hybrids, polydoxy, per/versions.[46] This is not to say that anything goes, but rather that our interpretations are always skewed by our contexts: we cannot escape the hermeneutic circle. Multiple versions, then, will always say more than any one version. At the same time, we must begin to look at our past traditions in ways that support our globalized, hybrid, evolving, nonsubstantial queer identities.

From this context and starting point, for instance, we can reread Vedic understandings of reincarnation along with the interrelatedness of all reality—past, present, and future—and begin to see how identities are empty of essence and always changing over time. If one thinks of time as "all together," then we can see reincarnation as a synchronic rather than diachronic process in Vedic and other traditions of South and Southeast Asia. From such a perspective, the multiple lives of the Buddha, for instance, can be understood together as "omnibodied, omnisexed, and omnigendered."[47] We are already, from this perspective, trans-species, cyborgs, elemental, and without origin or authenticity. From this place, very different understandings of internal and external, female and male, economies, and family values emerge. This, however, is all glossed over in the globalization of the Western rights-based capitalist politics of love, also known as family values.

Internally, here within the United States, the political-religious right proclaims family values to maintain an integral identity (heteromonogamy), while externally that same right imposes capitalism all over the globe. Such impositions, of course, rip other family structures apart, those that depend on large kinship networks, multiple marriages, alternative genders, common lands, same-sex relations, and companion species. This is all done internally under the guise of religious family values and externally under the guise of freedom and democracy (formerly under the guise of development). Such impositions over the face of the globe do indeed become the destruction of others' identities: they destroy both their understandings of religion and denigrate any concepts of nature that do not align with capitalist exploitation. These global impositions are, by all means, a war on others' meaning-making practices and ways of life. As such, I would argue that it is time we move beyond the capitalist politics of love and substantial notions of identity formation and move into identities that are emergent, ecoreligious, and planetary.

5 THE EMERGENCE OF ECORELIGIOUS IDENTITIES

> The power of the terms "woman" or "democracy" is not derived from their ability to describe adequately or comprehensively a political reality that already exists; on the contrary, the political signifier becomes politically efficacious by instituting and sustaining a set of connections as a political reality.
> JUDITH BUTLER, *Bodies That Matter*

Truths matter and that which you assent to becomes the world you work to create. The nature of truth is persuasion. What is true becomes the cocreation of the world that you live in. Certain systems support truths and thought habits that become natural or given. As meaning-making creatures, the two-fold bind we live in is constantly criticizing accepted truths while at the same time inevitably living by truths that are equally contentious. This is what it means to become a response-able meaning-making creature.

On the one hand, this book is only arguing that we are coconstructors of our reality: not creators ex nihilo as some total constructivists would have, but neither completely determined by some history, ideal, or material reality as Hegelians and Marxists might have. On the other hand, I am also arguing for more than the idea that we merely project our realities onto the world and then create them, for better and worse. What I have been arguing throughout the text thus far is that humans are meaning-making creatures. As such, we are also part of the rest of the natural world. This meaning that we make, however, is emergent from the ongoing process of nature naturing. Our identities, then, are never fixed by nature or transcendent meaning but rather emerge in ecosocial contexts in relationships with many earth others, both human and nonhuman. Such understandings of human identity, nature, and meaning have some radical implications, and I begin to explore some of those implications in this chapter.

What becomes of God, self, other, nature and other categories that create meaningful worlds? Rather than being ontological categories that describe any sort of reality, these all become permeable, ethical categories that shape the world around us. The way we make meaning, then, matters the world around us. In other words, it shapes our own bodies and other earth bodies around us, it cocreates with others the ecosocial worlds in which we live. Just as the substance-based metaphysic and meaning-making practice found in Western style metaphysics and its corresponding mechanistic model of science help to create the nature-cultures of the industrial revolution and all that entails, so now non-substance-based metaphysic and its corresponding interrelated and living models of science found in nonequilibrium thermodynamics, chaos and complexity, and other postmodern sciences are beginning to create new nature-cultures. Thus our meaning-making practices don't so much reflect metaphysics and ontologies as they do serve as regimes of truth in the Foucauldian sense. This chapter will explore our meaning-making practices as regimes of truth or what I call "technologies of meaning." I argue that two dominant regimes of truth are emerging and they are actually engulfing most of the meaning-making practices we have considered to be dominant in the form of the major world religions: that of economic globalization and a less popular but steadily increasing environmental planetarity.[1] Though both are truth regimes and cocreate ecoreligious realities around us, I argue here that the former truth regime, economic globalization, is creating a natural-cultural world in which differences give way to the sameness of the center. Such a creation, though just as natural or real as any alternative, does not create the kind of natural-cultural world that I find persuasive, for reasons I will discuss. Thus here in chapter 5 I begin my ethical arguments. These arguments are not based on any sort of foundational understanding of what is natural or unnatural or on what is right or wrong in some ultimate sense, but rather on what is aesthetically, socially, and ecologically persuasive. As a first order of persuasion, I end this chapter by arguing that we are always already constructed by transcultural ecoreligious identities. Such transcultural ecoreligious identities highlight the ways in which, as Judith Butler notes, we perform our identities, or, in other words, the ways in which our identities are performances of the ongoing process of nature-cultures nature-culturing. This performative understanding of ecoreligious identity coconstruction challenges much of what we think of as environmental ethics (chapter 6) and the very notion of human exceptionalism or species boundaries (chapter 7).

Technologies of Meaning

> "Human Becoming" expresses the idea that we are always in process, we are a becoming, and being human means that the journey is the reality—there may well be no final destination.
> PHILIP HEFNER, *Technology and Human Becoming*

In order to begin to jar us out of ways of thinking about meaning and to begin to understand humans as the dominant meaning-making creatures on the planet, though not the only meaningful creatures, it is necessary to shift our thinking about meaning from that of foundations in some sort of nature, God/Revelation or essential, original identity, and toward that of understanding meaning as *techne*. In other words, meaning, God, values, purpose, nature, and language are more about technologies of imagination than ontology or metaphysics. This is by no means a new way of thinking about meaning-making practices. For the Greeks *techne-logos* was simply words or discourse about an art form or craft.[2] It is not until much later in the course of Western history that *techne* gets separated into the mechanical and fine arts. Such a separation parallels the separation of any sort of inherent value/meaning from the material world and the placing of all value and meaning in a humanity created in God's image. More on this form of human exceptionalism will be discussed in chapter 7, but here I will just note that the mechanization of the material world during the Western scientific revolution was the result of one technology for understanding the world that eventually confused itself with the *only* technology. Heidegger, in his *The Question Concerning Technology*, understands this as the process by which the world is made into standing reserve.[3] We could also talk about this as the process by which one truth regime becomes normative or in terms of the dominant paradigm becoming confused with reality or in terms of misplaced concreteness or in many other ways that have been discussed throughout this book.[4] Here, however, I want to take a moment to offer a bit more of a subjective understanding of what understanding meaning-making practices and processes in terms of technologies might mean.

First of all, we have to begin to understand ourselves as natural-cultural, biohistorical, embodied thinking creatures. Second of all, we have to understand ourselves as part of a becoming process, as Hefner points out in the epigraph for this section. Such embeddedness means that we are not outside the becoming process of planetary evolution, nor are we outside the 13.7 billion year process of cosmic expansion. As such, we are both part of and actors in an ongoing process.[5]

Third, as actors in a network, or as embodied imaginings that are the result of previous histories of becoming worlds, ideas, bodies, and ecosystems, technology cannot possibly be understood as something that we use as mere instruments to create worlds around us. Rather, technology also shapes and forms our very subjective, bodily identities: we are included within the horizons of technologies of meaning rather than as the creators of those horizons. Heidegger again describes this as enframing: it involves the ways in which our very technologies and worldviews for becoming are the basis for our own subjectivities. We do not stand outside these technologies but rather they cocreate our very identities. He writes, "In the planetary imperialism of technologically organized man, the subjectivism of man attains its acme, from which point it will descend to the level of organized uniformity and there firmly establish itself. This uniformity becomes the surest instrument of total, i.e., technological rule over the earth. The modern freedom of subjectivity vanishes totally in the objectivity commensurate with it. Man cannot, of himself, abandon this destining of his modern essence or abolish it by fiat. But man can, as he thinks ahead [ponder the fact that this has not always been the case]."[6]

In other words, we are enframed by a modern technological mode of becoming that will turn even humans eventually into standing reserve, but we do have some capacity for thinking anew of alternate ways of becoming. Deleuze and Guattari talk about this existential experience in terms of "the fold."[7] If we exist on a single plane of existence, where there is no transcendent point, then the horizon of our thought is not like a transcendent movement pulling us beyond the here and now, but rather a fold in the plane of existence. As Catherine Keller notes about their understanding of the fold, this means that we live in an origami-like continuous creation: "The infinitely folded origami of the plane resembles in its virtuality, its potentiality, the 'milieu of milieus'. But then—as long as the Deleuzian virtual is not mistaken for a seamless unity—it contributes to the tehomic description of a matrix of possibilities."[8]

Enfolding is a difficult concept to grasp, but imagine that you are standing on the edge of the ocean looking out to the horizon. The last point beyond which you cannot see is where the ocean and sky meet in your vision. This is the metaphorical edge of your embodied existence. Now imagine that you were going out to that horizon, which of course would always be receding. Imagine that when you got say nine nautical miles out and reached the distance at which you saw the horizon from the shore, you would now be caught in a different embodied world where there was a new horizon and the distant shore. You have then folded a past horizon into the place of a new context for meaning making. You are not outside the context, but always an embodiment of the horizons and boundaries within which you live. From your perspective, you make meaning based upon the boundaries of your thinking, but this is just one perspective in a multiperspectival, planetary context that is always evolving or folding in upon itself. Multiperspectivalism will be discussed further in chapter 7, but here let it suffice to say that there is never a point at which you get beyond perspective or context to some ultimate horizon. You are always folding horizons and contexts into one another. As such, you yourself are an enfoldment of horizons. The practice of meaning making works in this origami-like way.

Meaning-making practices are shaped by histories and centuries of evolving life on the planet. As an individual you are born into such technologies of meaning, whether Western Christian, consumer/capitalist, Indonesian Muslim, Chinese communist, or any other combinations of meaning that mark the time and place into which you are born. Multiple technologies shape our embodiments and ways of thinking about the worlds in which we live. In this sense we are always already hybrids, cyborgs, and combinations of material-imagination.[9] To realize this does not mean that we are then able to create new worlds out of nothing, using technologies as mere instruments. Such an understanding would require that we could stand on an ultimate horizon or at some Archimedean point and recreate the world using technologies of meaning as mere instruments. Just as Butler's notion of performativity does not mean that we perform identities in ways we completely choose but rather that our subjectivities are performances regardless of whether they conform and subject themselves to normative ways of becoming or seek to per/vert those ways of becoming, so here we are always already captured by technologies or meaning-making practices. Just as equilibrium is death in a system of nonequilibrium thermodyanamics, so here a space of arrival, of salvation outside the mix, of paradise, or of perfect peace where "the lion sleeps with the lamb" is death. Such a liberative paradigm requires that there be an objective place toward which all values and lives can be judged.

As meaning-making creatures caught in and created by technologies of meaning, we are also shifters of those technologies, and it is possible to shift our ways of becoming in the world toward different technologies through the very space of the abject that is left out in any technology that tries to capture all of reality. In other words, it is the very leftover or remainder that questions any stabilization of identity or meaning.[10] This means that our freedom and agency lie in the collective remainders that return to haunt our technologies of meaning making: whether they be concepts of nature, concepts of religion, or identity constructions.[11]

Though there are many different technologies of meaning operating in any one context, here I want to discuss two emerging dominant meaning-making technologies that increasingly define the process of contemporary becoming: that of economic globalization and the emerging alternative of planetary becoming. Again, these are not pure categories, and there are many subdivisions and multiple other possibilities and even cross-pollinations among and between these two larger technologies. Nonetheless, I think it is possible to capture a lot of our contemporary contexts within a discussion of these two technologies.

Economic Globalization: Imposing Sameness

As I have mentioned elsewhere in the book, the basic distinction between globalization and planetarity used in this book is from the work of Gayatri Spivak.[12] For Spivak, globalization is the imposition of sameness over the entire face of the globe. That is the currently enforced technology of meaning; as such, it deserves further elaboration here. Many have discussed the problems with globalization, but I want to analyze its epistemology, sense of time, and ecosocial disparities. Before such an analysis, I must first proclaim that globalization is not all bad. In fact, I have written this book and you read it as heirs of a globalized context. I cherish the hybrid identities, mixtures of traditions and cultures, and privileged ability to move about the planet that is the result of globalization. Imagine how the Internet has led to revolutions or how queer people once isolated in small towns now have the capacity to reach out to a global network of queer peoples via the Internet. Globalization is a process and as such, like any other process, is not inherently good or evil. Rather, there are certain characteristics of the ways in which globalization is imposed upon the globe that create eco-social injustices for many. Let me begin, then, by just discussing three such characteristics.

THE GLOBAL EPISTEMOLOGY

In her book, *Sense of Place, Sense of Planet*, ecocritic Ursula Heise argues that we need to move from understanding the planet as the "little blue ball" spinning in space to the perspective of Google Earth.[13] The problem, as others have pointed out, is that the little blue ball image suggests an Archimedean standpoint from which the whole earth can be viewed together. Differences are glossed over at the expense of union and sameness. Such a global view of the earth does not reveal the various textures of terrain, class differences, or differences in health of planetary systems. Rather, as Bette Midler suggests in a song that she popularized, "From a distance, the world looks blue and green and no one is at war."[14] This distant view of the earth mimics the creation theology in which a monotheistic God creates the world out of a transcendent, objective space from nowhere in whose image humanity is created. In other words, it places human beings in a space over and against the rest of the natural world: as if we are little gods that can control and manage the little blue ball.

From this space of no-where, which is actually quite located in industrial, technological, and capitalist projects such as space exploration, which gave us the image of the little blue ball in the first place, the monological imposition of industrial-capitalist technology of meaning is imposed over the face of the globe. It may be called structural adjustment by some, development by others, and progress by still others, but the truth is that the forcing of Western industrial projects on the face of the planet helps to recreate the planet in the image of Western-style industrial capitalism. Liberation theologians, postcolonial scholars, ecofeminist scholars, and many others around the globe have described the horrors of this imposition, and I need not rehearse them here. The global imposition of sameness everywhere is based upon the mass accumulation of wealth and power by the few over the many. Such accumulations lead to a sense of speed and time that are literally outstripping the reproductive possibilities of many life-forms on the planet.

GLOBAL SPEED

In her book, *Globalization and Its Terrors*, Teresa Brennan identifies the primary problem with globalization as one of speed.[15] More will be said on the need for a nomadic environmental ethic in chapter 6, but Brennan's primary argument is that the horrors of globalization are not due to the fact that we all just need to develop a better monogamy of place and stay put. Rather, we need to focus on the pace at which we produce and make sure that it is not outstripping the reproductive capacities of the planet. As meaning-making creatures who are also caught in technologies of becoming, we can literally make meaning-full ways of existence that lead to the harm of many human and earth others. It is not that such ways of being are any less real than, say, living a preindustrial lifestyle off the land, but rather that they create different worlds and have different consequences. Again, Brennan's point is not that we find a more authentic or natural or God-given way of living in the planet, but that we pay attention to how our technologies of becoming affect human and earth others around the globe.

A technology of meaning, one based on the global view from nowhere, that suggests humans are indeed not emergent from the rest of the natural world apparently takes a lot of speed to reinforce. This speed comes at the cost of a lot of material and energy resources. The use of this volume of resources at such a rapid pace is only possible for a small portion of the human species, not to mention the other animals, plants, and ecosystems that make up the planet. As Bron Taylor suggests, we ought to follow the calories. It is all about the use and distribution of joules within the planetary community.[16]

The problem is that as countries begin to open up to globalization, this speed begins to take over all other technologies of meaning that might promote a slower pace. Getting into the system of moneymaking and opening up to global markets demands that cheap labor and cheap resources be spent in order to compete in the emerging global market.[17] In terms of daily life, this means that the majority of people work more and more for less and less. Furthermore, as will be discussed further in chapter 6, the erosion of the commons, in terms of actual land as well as social systems such as free health care and free education, leads many individuals with little to no choice but to join the pace of industrial-capitalist-economic globalization. We have to now pay for education with loans, we must get a job in order to have health insurance and pay for education loans, and there are fewer and fewer public lands on which we can depend for daily livelihood. In a very real sense, by the time most graduate from college in the United States, they already have a huge reserve of stored energy in the form of student loan debt. Thus, many students begin with an energy deficit that must be paid off through labor-energy. This example can be extrapolated to the deficit-producing modes of production that exist between the one-fifth world and four-fifths world or between human production and the rest of the natural world: we live as a result of energy deficits that must be paid by human and earth others (present and future). This leads to the third and final problem I see with the technology of globalization: ecosocial disparity.

The Global Mobiles and the Immobile Locals: Ecosocial Disparities

Zygmunt Bauman, in his book *Globalization*, talks about the process through which economic globalization creates two basic categories of peoples: the global mobiles and the immobile locals.[18] The global mobiles are those who, through the speed and wealth created by economic globalization, are "free to move about the globe." For these peoples, including myself, place does not matter. Many of us in the United States would fall into this category. The technologies that make meaning of our daily lives allow us to be placeless and in a sense reinforce the very idea that we are not *of* the rest of the natural world. In fact, this is reflected by the fact that anyone who lives in the United States, regardless of class, race, sex, and other distinctions, automatically lives at a pace that outstrips the Earth's reproductive capacity. One's ecological footprint, in other words, already requires more than the resources of one planet were everyone in the world to live as a person living in the United States.[19]

Once you are gripped by these technologies of high speed, in other words, it becomes a self-fulfilling prophecy in which your very independence is reinforced through your ability to move your body, ideas, and other bodies and ideas about the globe. Of course, what is backgrounded are all the bodies, materials, energies, and resources that such fast movement requires. Zygmunt Bauman refers to those who do not have such freedom as the immobile locals. For this group of people, mostly in the global south or what we might refer to as four-fifths of the worlds population, place is prison. Their cheap labor and resources are required for the wealth and speed of economic globalization, and this means they are often locked in place or forced to a new place as a result of environmental degradation and/or political instability. Furthermore, included among the immobile locals are other animals, plants, minerals, and ecosystems, understood to be resources for the global mobiles. These planetary others are seen as background to the global mobiles' imposition of progress over the face of the globe. The planet literally becomes a stage on which global identities are performed.

Given these ecosocial problems associated with the imposition of sameness over the globe and the technologies of meaning making linkeded to them, might we identify new technologies of meanings, new lines of imaginative flight, that would lead us toward different ways of becoming? Would it be possible to identify new technologies of becoming that promote the future of not only human becoming but also the becoming of multiple earth others? The metaphor of planetarity provides fodder for thinking about such new ways of becoming, and the rest of this book is about thinking with this metaphor of planetarity.

Environmental Planetarity: Becoming with Multiple Earth Others

Whereas globalization is the imposition of sameness over the face of the planet, planetarity is about connecting bodies and places of the planet through their differences—in recognition that our difference is what constitutes our very identities and localities.[20] In other words, the concept of planetarity takes a page from the ideas about identity formation found in queer theory. Rather than understanding our subjectivities or our humanity or our nationalities or our religions as being somehow formed in isolation and then brought into some sort of multicultural space of dialogue and exchange, it is through human and earth others that our subjectivities, humanities, nationalities, and religions are formed. At the very center of the self are multiple earth others. At the very center of what it means to be human is the multitude of nonhuman earth others. More on this will be discussed in the following section, but let me begin by offering the three counterpoints for a planetary technology of meaning: a planetary epistemology, a planetary understanding of time, and the planetary promise to address ecosocial injustices.

PLANETARY KNOWING AND BECOMING

One of the major differences between a global and planetary epistemology is that planetary knowing always includes becoming. In other words, planetary knowing is contextual; there is no objective space of transcendence from which everything else can be recapitulated into the image of the objective space. Catherine Keller writes about this as "grounds" for knowing rather than "foundations."[21] Such a postfoundational epistemology recognizes that we are always already creatures with text (contextual). As contextual creatures, we are not left with relativity. Rather, relativity and universality are two sides of the same coin.[22] Relativity means that the only context that matters is my own (and thus no other context matters) and universality means again that no other context matters because one mistakes her own context for the only context. Indeed there are common grounds we can agree on: we are on a planet, we are all mammals, we breathe the air, and gravity seems to apply to us all. None of these examples is true for all times and all places (i.e., universally); we were not always on this planet nor was this planet always here; we are mammals now, but perhaps we are evolving beyond ourselves (see chapter 7); we breathe the air, but the mixture of the air that we breathe has not always been constant and may not always remain; and gravity applies to most things, except every point of the universe seems to be rushing away or expanding so that gravity does not hold at the edges of the universe.

Our common grounds, much like the ground below our feet, shake, move, and shift. As Thomas Tweed notes in his book, *Crossing and Dwelling*, our sense of home and place is only a dwelling, "for a time."[23] This suggests an epistemology and ontology of plate tectonics. It is only from these grounds that we can begin to greet and dialogue with the human and earth others that make up our very contexts. Rather than approaching human and earth others with a transcendent monological understanding of the world, we might from this place approach human and earth others through a dialogic form of knowing.[24] Such a dialogic model recognizes that knowledge is partial, located and that we live in a multiperspectival planetary reality. As such, my knowledge will never exhaust the meaning of a human or earth other. I can never exhaust the knowledge of you, nor of a tree or a dog or an ocean or a galaxy. In this sense, our unique subjective experience of the world is universal in that our own unique subjectivities will never be recreated exactly the same again. This multiperspectival, contextual, planetary epistemology takes on a different understanding of time as well.

PLANETARY TIME: REVERBERATIONS

Whereas the time of the global logic of domination is that of an increasing speed and an imagined linear direction that develops places out of poverty and into technocapitalism, planetary time with its multiperspectivalism is much different. The linear, progressive understanding of time goes hand in hand with the monological imposition of sameness over tface of the globe.[25] Furthermore, it could be associated with Zygmunt Bauman's notion of "liquid modernity" and the love of the new.[26] Again, whether we speak of the universal imposition of sameness everywhere in the progressive understanding of time or relative egoistic fragmentation of perspectives in the love of the new, contexts are denied.

Planetary time, on the other hand, must listen to the multiperspectival bodies and realities of the planet. Time here is much less about progress, or even cycles, than it is about constant reverberations. Contextual time, much like the origami folds of an immanent understanding of reality, is constituted by the past, present, and future possibilities of the voices and bodies of a given context. That is, from the present time we look toward the past, but only from our present context, which is also always defined by our hopes, dreams, and visions of future becomings. Who is to say whether the voice is coming from the future or the past? If you have ever been sitting between two buildings, or in a canyon, and hear a noise that seems as if it is coming from in front of you but is really coming from behind you, then you will get what I mean by this experience. The present reverberates in much the same way between past and future, except the mathematics and physics of sound do not offer such precision in discernment when it comes to the realm of the present experience of past and future. Is what we interpret as history where we want to go? Is what we interpret as history rooted in where we have come from? Both versions of history, those relying on origins and ends, are put into question.

In a reverberating present—marked by infinite regress in terms of origin, the context that gives rise and shape to the sound and openness as the sound travels outward and cannot be contained—our epistemologies should be geared toward dialogue, toward the recognition of multiple perspectives. There is no solid past and future, but the becoming present decides how we interpret the past in accord with what visions we desire for the future. This is the bold reality of the *post-* in the *postmodern*: without knowing where we are going we cannot project certainty into the past. This is the Deleuzian-Guattarian rhizomatic ontology-epistemology that recognizes contextual ethical responses without justification in origins and ends. The "reverb" of *re-verb-eration* implies practicing reiterations, which suggests the ability to respond. Response-able. Responsible. Responsibility for the knowledge that we are always already with text, and taking responsibility for that text is the place where our thinking and meaning making begins.

Response-able Planetary Creatures

Far from suggesting here that ecosocial problems will ever be resolved, what I do want to offer is a method for keeping our focus on the multiple, emergent planetary contexts of which we are a part. As Latour notes, religion is not about taking us far away, but about keeping us focused on everyday contexts.[27] A linear, progressive narrative such as is found in the global technology of meaning making pushes our thinking and our responses away toward a future that is much more the projection of our present than we care to think. In other words, from within the global mobile understanding I have discussed, the idea of progress is really about making sure that the apparent freedom presently experienced by the global mobiles is maintained indefinitely into the future. Ecosocial ills become, then, a matter of enforcing a very located, elite meaning over the face of the planet without any attention to how the past and present conditions of various locations make such technocapitalistic speeds impossible. Solutions to problems are always found in more growth, development, or technological progress. In this sense, we become less able to respond to multiple, hybrid, and emergent planetary contexts: the pathway forward is one.

From within a planetary technology of meaning making where time is more about reverberations and the continual strum of the strings, our attention is focused on how we strum the strings of the planet and what types of reverberations are sounded. It is the strum of the strings, the vibrations, that gives us our context for the next moment of becoming. We are always already trans, in-between, hybrid, and made up of multiple earth others.

Transcultural Ecoreligious Identities

The embodiment of trans locations captures our bodily and lived realities much more than any model based on the original, orthodox, pure, singular, or universal can offer. In other words, the planetary reality is one in which we are hybrid, cocreated, and always already in the mix with multiple earth others. As such, and given that we are meaning-making creatures *of* the planet (again, not the only meaning-making creatures, but this is our defining quality), I would propose a transcultural and ecoreligious understanding of our various embodiments. Such an embodiment relies much more on infinite regress than any sort of foundationalism in defining our subject locations. In this brief section, then, I want to bring to bear what the language of queer theory and postfoundational understandings of both scientific claims about nature and religious claims about reality might mean for planetary identities. At least three realities mark these transcultural ecoreligious identities: our subject positions are always already multiple, our embodied thinking is always in-between, and our actions, our agency, is always shared by multiple planetary others.

The Multiple Subject

As discussed in the last chapter, understandings of subjectivity, especially in Western traditions, have been based upon foundational and/or substantial ways of thinking. In other words, there is something about our subjectivity that is unchanging, eternal, that lives on in the face of all our life changes. Such subjectivities understand the self as a unified, solid subject. This subject is at the heart of the Lockean liberal understanding of subjectivity, which in turn is at the heart of capitalism (private property and labor) and our legal system (individual human rights). If, however, we begin to think in postfoundational ways, our subjectivities become multiple sites of resistance for the imposition of monological thinking upon human and earth others.

As suggested by queer theory and deconstructive thought, when we define our very subjectivity we will always be leaving something out as remainder, as that which is abject and returns to haunt the very definition of the self we mistakenly equate with a stable identity. It is this remainder that is always necessary and destabilizing for self-definitions. It is this haunting remainder of the self that, much like a trickster figure, returns to blur the boundaries between other and self, nonhuman and human, female and male, and all other categories of identity formation. These blurry boundaries are precisely what enable us to continue to grow and change in both literal and figurative ways. As Terrence Deacon notes, the "I" may be much more like the absence of the center of the wagon wheel than what is identifiable in presence. It may be that what is abject, what haunts, is the very source of our own unique agencies and identities.[28]

Furthermore, our subjectivities are multiple in that we are made up of many human and earth others: histories, societies, actions, earth, air, water, fire, other molecules, other plants and animals. We are quite literally not the creators ex nihilo of our own identities, but we are created by multiple earth others. In a very real sense, we cannot cut off our understandings of the self from the whole 13.7-billion-year process of cosmic expansion and 4.5-billion-year process of geoevolution that led up to the very emergence of hominoids and eventually *Homo sapiens*. Though we do make a heuristic cut at some point in order to interpret our lives (which we have to do as meaning-making creatures), we must always remember that these interpretive cuts ought never be confused with reality in itself. Instead, we are always already multiplicity, coming together on a moment-by-moment basis in subject positions that can never exhaust the full reality of our embodied becomings.[29] From this perspective, perhaps planetary technologies of becoming will encourage us to think with earth others—and think with the in-between rather than as isolated thinking things.

Thinking In-Between

Timothy Morton, in his book *The Ecological Thought*, discusses the ways in which we make these hermeneutical cuts and equate them with reality. He suggests that perhaps we ought to begin thinking in terms of hyperobjects and hypersubjects.[30] What he means by this is that we are all collections of strangers: even our own subjectivities are made up of multiple strangers. "The ecological thought imagines a multitude of entangled strange strangers."[31] Donna Haraway speaks about this reality in terms of "companion species."[32] Finally, Deleuze and Guattari speak of this type of multiple subjectivity as bodies without organs.[33] The point of all three of these very useful metaphors is to help us begin to think of ourselves as in-between. Such thinking is not new: postcolonial and feminist thought have especially been arguing for relational and multiple understandings of the self for decades. What perhaps is new in terms of these metaphors is their indebtedness to what I have been describing as planetary ways of thinking. If our subjective understanding is always already multiple and includes human and earth others, then our thinking ought also to include these multiple earth others.

Thinking in-between from a planetary perspective means breaking out of our instrumental ways of thinking about the rest of the earth (chapter 6), as well as out of our anthropocentric understanding of conscious thought (chapter 7). Humans, in other words, as meaning-making creatures made up of many unique earth bodies (and also making up other earth bodies) are not like little, isolated, Cartesian thinking-things opening the world to a reasonable understanding. Rather, our thinking is more like trees transpiring, weather patterns forming, plate tectonics shifting, or birds migrating. It is something that we can't help but do, for starters. But, even more important, our thinking is impossible without the multiple subjects and lines of flight that make up our own bodies and minds. Our minds are being played by evolutionary processes and by the multiple earth others that make up the contents and contexts of our thoughts. The Cartesian self is but an illusion and would have no content for thinking were it a reality. Of course, we can cocreate the technologies that support this understanding of the Lockean Cartesian self, but the question raised is at what cost do we create such individualities. From a planetary identity, it is on the interstice or the boundary or the borderlands where self and other exchange that our thinking emerges. Just as our thinking is in-between and shared, so are our actions, which brings me to the concept of shared agency.

Shared Agency

The idea of a multiple subjectivity challenges notions of a unified Cartesian thinking thing just as it challenges the idea of the commanding ghost in the machine. If we are multiple, and if imagination-material, mind-brain, or spirit-matter are always already involved in one another, then the concept of a unified commander ordering individual central nervous systems around is also challenged. Again, emergence theory is trying, at least in part, to think a third way between monism and dualism. Accordingly, it navigates between a complete top-down and a complete bottom-up version of causality. Top-down causality suggests that there is some sort of platonic ordering form: mind over matter, language constructs reality. Bottom-up causality alone suggests that it's all in the genes or it's all in the neurons. In other words, both top-down and bottom-up approaches to reality reduce all reality to one side of the material-ideal spectrum and reduce causality to efficient causality alone.

As mentioned in previous chapters, rather than understand the sequences that lead to the emergence of new life-forms and phenomena as strictly the result of efficient causality, we need also to understand phenomena and bodies in terms of formal and final causality.[34] Formal and final causality have to do with the habits, supported by institutions and histories of thinking, that shape our becoming bodies into certain ways of becoming. As such, agency is not just a matter of individual choice, but rather it has always already to do with the contexts and circumstances under which individuals are born and live. This has radical implications for our legal systems.

Though many will agree that histories of racism, classism, sexism, and heterosexism help to give certain peoples "unearned privileges" in the world,[35] few find it realistic to then address the formal and final causes that bring about that unearned privilege. If we only focus on efficient causality, then individual humans will always be held responsible and the larger economic, legal, health care, educational, and other systemic reforms that need to take place will be ignored. More will be said about such changes in the following chapters, but here I will provide one example: the prison system in the United States.

It is no secret to most readers that incarceration and crime and punishment in the U.S. (and, increasingly, globally) are focused on the individual responsibility and agency of those involved in the immediate circumstances of the crime. As such, histories of classism and racism are not taken into account when one is charged with a violent or nonviolent crime. This situation leads to one in which there are a disproportionate number of poor people and, especially, people of color in our prison systems. In fact, black males in the United States have a higher probability of going to jail than graduating from college.[36] Furthermore, all legal studies suggest that the use of the death penalty, among other harsh penalties, is determined by racially motivated causes (whether overt or covert).[37] Thus there are many years of habits (racism and classism, for example) that have to a great extent helped cause certain people to be the targets and victims of the legal system more than others. If we look only at efficient causality, these formal and final causes will never be addressed and will only serve to perpetuate the racism and classism already in place.

From this perspective, we need to radically overhaul our legal system. We need to address histories of oppressions through reparations and affirmative actions. Given that the death penalty is applied according to race/class, and that agency is shared, we need a moratorium on such practices. We should review the racial history of criminalizing drugs and the violent behaviors that the so-called war on drugs perpetuate throughout Central and South America and perhaps decriminalize some of these drugs if not outright legalize them. In place of these technologies of policing individuals, we can focus our attention on education, physical and mental health programs, and retooling our economic and legal institutions in such a way that they help create new habits for bodily becomings toward a less racist, sexist, classist, and heterosexist future.

With shared agency, we are neither completely free from our contexts, histories, and material becomings, nor are we completely determined by them. One of my friends and colleagues who writes about emergence theory, explained it well. She said something along the lines of the following: "I used to think 'I' was in complete control, now 'I' like to sit back and wonder what 'I' will do next." This for me is reminiscent of the skylike mind in Buddhism in which you let your thoughts and feelings flow past like clouds. You realize you are not in complete control, yet you also have some ability to determine how you will respond to your thoughts and emotions. It is here in this ability to respond that I find the connection between this emergent theory of identity, thought, and agency and the concept of performativity.

Performing Meaning: Taking on the Abject Toward Planetary Identities

> Vital materiality better captures an "alien" quality of our own flesh, and in so doing reminds humans of the very *radical* character of the (fractious) kinship between the humans and the nonhuman.
> JANE BENNETT, *Vibrant Matter*

The concept of performativity developed by Judith Butler helps us to navigate these newly emerging planetary identities. If we are emergent, meaning-making creatures with shared thought and agency, then performing and per/versions are the places where our capacity for change resides.[38] In other words, the biological and historical lines of flight or habits that shape our ever becoming embodiments are not prisons. We are not fully determined by our contexts, but we must subject ourselves to these contexts in order to have any sort of subjectivity.[39] At the same time, such subjection always already creates the abject, and it is in "taking on the abject" that we can perform per/versions of meaning. This "taking on" means both "strapping on," or trying out, the abject identity and taking it into consideration as it confronts and challenges our own identities.[40] From this perspective, the abject is revealed as the place of power-with and thus precisely not terror or power over. In this final section of chapter 5, then, I want to explore how it is that taking on the abject can move us toward a multiperspectival planetary identity.

The abject, again, is that which gets excluded from the process of identity formation but returns to haunt the very boundaries of a given identity. It is this haunting that enables one's identity to remain open to the other. Much like Latour's actor-network theory (discussed briefly earlier), our identities are per/formed with multiple per/versions. Our identities, including our biologies and histories, are like scripts that can be interpreted in multiple ways, similar to the way in which modifications/mutations are made in the process of gene transcription. In any given biohistorical context, certain parts of a script will be more important than others. Furthermore, the script can be rewritten, modified, and changed over time so that it responds to the evolving contexts. How many times has *Romeo and Juliet* been recast and rewritten for contemporary audiences? Similarly, this is how various religions are performed, renegotiated, and evolve over time. It is the network of actors and their connection with the audience that makes or breaks any given manifestation of the characters in the script: for us, this would include human and earth others.

The point is that, for any script to be performed, there must be flexibility for the characters to adapt to the context and connect with the audience. At the same time, the actors must come together and perform the script, otherwise no meaningful exchange would be possible. Again, in the language of emergence theory, it is the evolutionary constraints that make our lives, freedom, and further evolution possible, much as scripts make the ongoing production of plays and movies possible. Such is the case with our identities: the process of becoming a subject requires us to subject ourselves to a certain way of being in the world. However, that very subjection will always leave out something other from us. The violence of collapsing any given manifestation or performance of subjectivity into an essential identity causes the abject portions of our identities to be abandoned, and the self is thus cut off from that which is other. Rather than participate in this cutting, what if we took on the abject, holding it lightly in our self-understanding, allowing it to haunt our identity? Such openness to the abject/other might help us recognize that it is this very other that allows us to remain open and grow. Without this cascade of others, then, our subjective journey would have no horizon toward which to journey. The goal is not grasping and consuming the abject/other, but taking it on in its full otherness. Such taking on leads to a self-understanding that always already includes a multiplicity of human and earth others.

Here, at the end of this chapter on identity construction, we begin to see how a multiperspectival planetary understanding of meaning making might challenge the boundaries of self and other. This has huge implications for our legal, educational, and economic systems. In the two remaining chapters I explore what such multiperspectival, planetary identities might mean for environmental ethics and transhuman identities. As discussed in previous chapters, the notion of place in environmental ethics helps keep us locked into our identity constructions in ways that may not be helpful; hence in chapter 6 I call for a nomadic environmental ethic. Then, in chapter 7, I challenge the seemingly impermeable boundary between human and other planetary subjects, suggesting that if we take planetary technologies of meaning seriously, we must begin to imagine our own lives evolving beyond the species boundary.

6 DEVELOPING PLANETARY ENVIRONMENTAL ETHICS

A Nomadic Polyamory of Place

> A world marked by multiple gods who do not place human welfare high on their list bears an uncanny affinity to a world of becoming composed of multiple, open force-fields of numerous types.
> WILLIAM CONNOLLY, *A World of Becoming*

According to ecocritic Ursula Heise, "The crucial insights of the last twenty years of cultural theory into the ways local and national identities depend on excluded others, how these rely on but often deny their own hybrid mixtures with other places and cultures, and in what ways real and imagined travel to other places shapes self-definition" have not really been taken up in Western forms of environmental discourse.[1] Much of ecological and religious thinking is tied up in securing home, community, place, identity, value, and meaning. This is juxtaposed with the globalized, postworlds in which we live, where such stability, arrival, and homecoming is not possible, nor perhaps even desirable. Might such a desire for stability, place, returning home, or "getting back to" actually exacerbate the ecological and social problems we face today if a robust understanding of global movement does not accompany placed-based thinking? These ethical, epistemological, and ontological questions are at the heart of this chapter, in which I examine the concept of place at three different levels: what I am calling the level of epistem-ontological or slippery slopes, the ethical/religious level or common grounds, and the anthropological/political level, or uncharted territories.[2] I suggest what is needed is both a polytheistic nomadic environmental ethic and what I am calling a polyamory of place. Whereas a monotheistic nomadic ethic results in globalizing sameness, a polytheistic nomadic ethic links the planet through differences. Whereas an environmental ethic derived from place-based thinking calls for deep love and monogamy with a single place, a polyamory of place calls for love of many places as part of a larger, planetary community. Thus our epistemologies or ways of knowing the planet become polydox and our ontology, or ground of life, becomes multiple. An ecology of movement rather than location becomes the key to addressing contemporary ecological ills. Because our religious and scientific knowledge of the world is fluid, and our identities are fluid, in this chapter I begin to articulate ethical guidelines for planetary action that are fluid, changing, and context dependent. Such an ethic understands the human need for movement and seeks an "alternative hedonism" in environmental ethics rather than merely apocalyptic rhetoric that forces us to recede into an aesthetic of singular place.[3] As Plumwood reminds us, "The very concept of a singular homeplace or 'our place' is problematized by the dissociation and dematerialization that permeate the global economy and culture."[4]

Planetary Flows: A Context of Movement

The human world is made up now of mostly urban peoples, and this means that one hallmark of contemporary anthropology is movement. Most people in urbanized/industrialized worlds could readily recount their personal narratives of mobility, privileged or otherwise. My own has taken me from Stuttgart, Arkansas to Hot Springs, Arkansas to Little Rock, Arkansas to Conway, Arkansas to Budapest, Hungary to Nashville, Tennessee to the San Francisco Bay Area, to Miami, Florida, Indonesia, India, and, briefly, to several other distant places around the planet along the way. Yet, in spite of this recognition and partial acceptance of movement—movement that enables academicians to do the very work that we do—I wager that I am not alone in longing, at times, for an idealized place/home. From whence does this desire spring?

Movement can be addressed at many levels: think of all the forced migrations and refugees due to war, famine, environmental problems, economic inequity, general political unrest, and genocide. Or think of the historical and contemporary movement of religions around the globe through missionaries and through media. Or think of the movement of pollen in the gulf stream or waters around the ocean, or plate tectonics, or migratory birds, or human-introduced species of plants and animals. Or think of genetic adaptations, neurochemical transmissions, the flow of energy, transpiration, evolution, and even cosmic expansion. This is a planet and universe on the move. Yet the priority of an idealized place is revealed when species are problematically termed exotic and, sometimes even worse, invasive exotic or when human immigrants are referred to as alien or illegal. What, in light of these ever changing realities, can be preserved by a place-based analysis and how might such an analysis be complemented by an analysis of movement, or a nomadic understanding of environmental ethics? What does it *mean* to be but one species in a planet on the move? The rest of this chapter will be devoted to answering these questions.

But, first, a brief precautionary statement is needed. I am neither condemning place nor claiming that all place-based theories are parochial. In fact, as many essays in Sigurd Bergmann's and Tore Sager's edited volume, The Ethics of Mobilities, reveal, place can also include analysis of movement.[5] Bergmann writes, "Technical and social processes of acceleration become more and more insensitive and therefore destructive of ecological processes, where the speed of biological life cycles and development do not follow the principle of constant acceleration."[6] Yet the book is far from a cry to return to place. He goes on to say, "Feeling at home (*Beheimatung*) does not necessarily mean that one simply has to stay at home; it also must include the motion of leaving a place and returning to it."[7] Movement, then, becomes characteristic of home and place, and the task here is to try to articulate "the significant notion that place itself is nomadic."[8] Thus I am not valorizing movement. Certainly the speed at which we move using jet fuel rather than trains and boats may not be sustainable. In fact, as Teresa Brennan, among many others, has argued, in our current "consumptive mode of production, the artificial space-time of speed (space for short) takes the place of generational time."[9] The speed at which the globalatinization of industrialized free-market systems move is quite literally outstripping planetary places.[10] So this is not a valorization of the pace of progress. Rather, what I am suggesting is that the reaction to the movement characteristic of the globalized world does not have to be a romantic return to place, which in many cases is an ostrich response to global ecological ills. What I am offering here is a constructive critique of some of the implications of place-based thinking in an evolving, planetary context. Furthermore, I want to affirm that which is beneficial about movement and provide, perhaps, an erotic lure to think planetarily about the ways in which we move. As Juliet Solomon notes, "Movement and mobility" can act as "generators of flow experiences, as creators of personal space, as suppliers of hope, as expanders of our worldview: these are important functions which need to exist."[11]

What does place look like if we take seriously the current reality of a planet that will probably never again (if it ever did) have isolated localities? Somehow the commandments, "don't fly, stay put," or "keep the landscape free of human influence or exotic species," or "if you don't know every species of every plant where you are, the watershed you drink from, and/or the entire ecological history of your place, then you are a bad environmentalist," just don't persuade people to jump on the environmental-bandwagon. These platitudes evoke shame in a world that is filled with other options. It is no wonder that apocalyptic or shame-based environmental discourse does not take hold in a way that forces some sort of revolution. These types of discourse take a page from antiquated religious commandments that rely on a transcendent other that is keeping score. Furthermore, this type of discourse comes at a time when more and more people are questioning the transcendent claims of their own religions. Religious narratives that valorize the past and make apocalyptic the future-present, usually by turning the future into an idealized version of the past, no longer make sense amid the mixture of religious ideas resulting from globalization.[12] If religious and environmental narratives are going to continue to persuade ever new generations—who have never known anything but this globalized context—into living toward "better" visions of the future, then we may need to adapt our tropes and rhetoric toward these ever changing contexts. With this brief caveat, I now move to three levels of a nomadic understanding of environmental ethics.

Slippery Slopes: The Epistem-ontology of Planetary Ethics

I suspect that many peoples attracted to some sort of environmentally sound vision of life would love nothing more than to subscribe to some sort of place-based lifestyle. Whether urban (I could get lost in the San Francisco Bay area for the rest of my life) or rural family-based homesteading or more intentional community based, there is a certain sense of longing for a specific place. What if that very sense of longing is the result of an idealized and internalized understanding of arrival? What if the very places we so long to inhabit are what prevent us from developing a viable planetary community, and polyamory of places, through reifying our sense of meaning, belonging, and ethical concern in such a way that we become overly partial to the local and ineffective at addressing global concerns and even recognizing global assemblages? I argue here that monogamous love of place can be quite detrimental to planetary flows of life.

An attachment to place coupled with a reality that literally moves us around the planet can lead to a certain dismissal of place for fear of the pain of having to leave it. In *Apocalypse Now and Then* Catherine Keller writes of the tendency toward apocalyptic thinking when we move from place to place. In order to convince oneself that one wants to leave, we tend to demonize the place we are leaving.[13] Is this a result of deep love for the place and a method of cutting ourselves away from a place? Is this apocalypticism of place the result of not having a healthy, ecospirituality of moving through places as an embodied agent, but rather of one that tells us we should stay put? Probably it is a little of both. As Slavoj Žižek notes, "Although ecologists are all the time demanding that we change our way of life, underlying this demand is its opposite, a deep distrust of change, of development, of progress and [of the idea that] every radical change can have the unintended consequences of triggering a catastrophe."[14] This focus on place, on return, is very much related to the fear of the other that so many postmodern scholars have highlighted. If postmoderns such as Gilles Deleuze and Felix Guattari, Jacques Derrida, and Homi Bhabha (in their diverse ways) are correct, it is only through others that our own identities are formed. Thus to perform apocalypse on a specific place may also be to reify our own isolated identities and our own sense of place.

Environmentalisms in their Western forms have become so associated with place, especially wild places, that they stumble and stutter when trying to address the planetary realities of the *glocal*. Part of the problem is that Western environmentalism to a large extent relies on foundational claims about a nature that can be regenerated, restored, preserved, and conserved. In order to perpetuate the belief in this sort of nature, one must continue to perform place-based identities or else one must change one's views of nature. One must connect with place-based nature in order to keep the possibility of this genuine arrival alive. In other words, Western environmentalism is geared toward arrival: at a place, an identity, or a so-called genuine relationship with nature. As is well known, in deconstructionist thought the idea of arrival as a goal is problematic. It is problematic in part because arrival always involves a backgrounded exclusion. That is, arrival always includes an ignoring or erasure of the ever-changing, unstable, ways in which selves-and-others constantly negotiate the permeated boundaries of identities. Furthermore, Val Plumwood also reminds us that any time we buy into such idealized places we forget the "shadow places" on which our construction of place depends. These idealized places enable us to become "more and more out of touch with the *material conditions (including ecological conditions) that support or enable our lives.*"[15] That is, a concept of isolated place enables us to ignore the many different locations, peoples, and life-forms that are effected by our ecological footprint. Ignoring these messy negotiations can then cause violence to human and earth others as identities are forced to fit a one-size-fits-all mentality and other identities are ignored all together.

Teresa Brennan discusses the psychological production of this longing for arrival in what she (following Melanie Klein's development of Freud's psychoanalytic theory) has identified as the "foundational fantasy." She writes, "the foundational fantasy is the means whereby the human being comes to conceive itself as the source of all intelligence and agency."[16] She argues that the foundational fantasy is that original denial of dependence one makes when one forms the understanding of self as the isolated individual. Symbolically, in patriarchal religions, "the denial of creativity and dependence begins here with the denial of the mother" and the affirmation of a male creator god of all life out of nothing![17] This is the metaphorical image of independence par excellence. Much like Plumwood's understanding of the backgrounding of others in the name of the self-sufficient individual, Brennan suggests this type of fantasy enables the commodification of the globe that takes place in economic globalization and the subsequent creation of individuals as primarily consumers. The problem is that our identity boundaries are fluid, we are interrelated with other people and the rest of the natural world, and our categories of meaning and thinking can thus never capture the reality of things in full. This level of uncertainty in epistemology (knowing) and ontology (being, but here I mean in a more broad way the realities we believe and live in) is not easy to accept, especially for those living in the one-fifth world.[18] The more economic power one has, the greater amount of centralization she experiences, which in turn maintains the illusion of centralization or the foundational fantasy: commodities are brought from around the globe to a central location in the one-fifth world. Zygmunt Bauman, as we have seen, refers to this phenomenon as the production of global mobiles and immobile locals within the process of globalization.[19] The closer one is to economic power, the more one benefits from the globalization of the economy and the further one is from the ecological and social consequences of his actions. It is the four-fifths world peoples that feel more strongly the effects of the spread of free-market capitalism and hyperconsumption across the face of the globe. Though these categories are, of course, fluid, the dynamic Bauman identifies is, I think, accurate. Further, "the price [of the illusion of independence] is higher the greater the extent of centralization and globalization, the more distance has to be covered at a higher speed, the more energy consumed, and the less the [regenerative energy is] returned to nature."[20] In other words, it takes more and more technological and communications speed to maintain the illusion of independence the more we live in a globalized world. This type of speed is, according to Brennan, outstripping the (re)generational time of nature. "Spatial centralization creates energy demands and an energy field which can only sustain itself by extracting surplus value from nature."[21] Perhaps, as has been discussed elsewhere by many (including myself), the foundational fantasy of an omni-God that creates ex nihilo is part of the very reason that place and arrival are valorized over movement through places and journey.[22]

In understandings of place, arrival is valorized over taking leave or movement. Could it be that nonarrival, movement, nomadism, and change are actually better ways of understanding our contemporary globalized planet? Even when we seek out and create places in an ecosystem, region, or nation, we do this by backgrounding our own ecological/agricultural/economic nomadism, which affects the globe in very material/matter-real ways via the processes of economic globalization. In a way, we can't wrap our minds around never arriving, so we fool ourselves into thinking we stay put, all the while using technologies that bring the rest of the world to us. We sit around "waiting for Godot," in this case, waiting for the confirmation or justification of our independence. At least in this way the *illusion* of place, nation, region, is maintained. But, the cost is the destruction of other places. In other words, I can sit at home, type up an editorial about the destruction of the Gulf of Mexico due to the BP oil spill, place that story on a blog, and join the global conversation about it while never reflecting on the energy demands of the technologies that go into, say, the energy needed to maintain the grid of the World Wide Web or the toxins that go into making a hard drive. Both depend on oil—though they don't have to. Furthermore, I might only attack BP in that editorial and never acknowledge the fact that BP awarded UC Berkeley with one of the largest grants to explore alternative energies (to the tune of $400,000,000).[23] This grant even launched a new image campaign that would change British Petroleum into Beyond Petroleum. All this is easily lost under an illusion of spatial and temporal place that backgrounds the complex flows that go into any given situation, identity, or place. Might an ethic of spatial and temporal movement highlight the complexities of the situation and circumvent the desire to remain located as an individual in an isolated psychological, social, and ecological place? This illusion of place is tied to the illusion of foundational thinking. Both rely on stable identities or concepts. Both background that which does not fit into a given identity or concept. Thus the critiques of foundational thinking and the violence that ensues from foundational thinking may also be fruitful when applied to place.

A move from foundational thoughts/fantasies of nature and self toward that of infinite regress, or slippery slopes, is something to be at least entertained. Fear surrounding the idea of infinite regress in knowledge claims and in identity formation is belied by the very naming of an argument a slippery slope argument. This terminology is often used by environmentalists and others who abide by the precautionary principle. The precautionary principle favors maintaining the assumption of a stable, foundational nature, and movement or change is only justified when an unrealistic amount of certainty presents itself. Here environmentalists and those who think the environmental crisis is a myth may be guilty of the same fear: that of change. For environmentalists, changes and developments in technology or changes in nature are to be feared; for those who are anti-environmentalists, changes in everyday habits are to be feared. Making a change admits fallibility or at least uncertainty and slipping on the slopes of uncertainty may quickly snowball into tragedy. It is precisely this reliance upon certainty in thought that is made problematic by critical discourses: in other words, thought *cannot* enclose, capture, or found securely evolving realities. Life is an emergent process that is open, uncertain, and always a bit messy. Claims about nature are always made on slippery slopes. There is no bottom to nature, nor top for that matter: there are multiple levels, or folds, to use Deleuzian language, from which one can analyze and experience nature, including the physical, chemical, biological, ecological, social, cultural, and religious levels. And each of these levels is open toward a becoming, emergent future.

When we make claims about nature we always already do so on slippery slopes: there is then an infinite regress when making claims about nature. Thus we make a decision to define it this or that way; we construct slippery descriptions and work toward persuading others of them. As others slide into our understanding, and we into theirs, we realize that we can find common grounds, but also that there are multiple ways in which we might become. As Deleuze and Guattari note, "If thought may unfold across a thousand plateaus, there are a thousand ecologies that would unfold within those plateaus, a thousand ways of attempting to create a new collectivity and a new earth."[24] The point is not that we can never define places, nature, or our own identities, but that these descriptions are always already nomadic, on the go, unfolding, deterritorializing, and reterritorializing. As Timothy Morton notes in *Ecology Without Nature* and Bruno Latour in *The Politics of Nature*, it may be this very stable, place-based, foundational understanding of nature that gets us into trouble in the first place, precisely because it backgrounds the process of the construction of nature and thus takes nature out of political processes.[25] When nature is taken out of the political process of negotiation, it only becomes a servant to the reification of one's identity. Similarly, Spinoza's concept of nature naturing (rather than that of nature natured) suggests that nature is an open, evolving process and therefore cannot be trapped or confined to our concepts of nature. Recognizing this, perhaps we can begin to take responsibility for our definitions and redefine ourselves, our places, and the planetary community in terms of ever new and more expansive common grounds from which to address social and ecological ills.

The Ethical/Religious Level: Common Grounds

Mary Evelyn Tucker, coeditor of the Harvard series on *Religions of the World and Ecology* and cofounder and director of the Forum on Religion and Ecology writes: "The common *ground* is the Earth itself as an expression of numinous creativity, a matrix of *mystery*, and a locus for encountering the sacred. This common ground of mystery is in danger of being *blindly* wasted. It can be said, then, that the environmental crisis may disclose not only the common ground of the mystery of the Earth itself, but also the higher ground *beyond differences* in the search for the common good to promote the flourishing of life. In this effort common ground and common good are joined."[26] This captures well a few things about developing a nomadic understanding of planetary identities. First, it highlights the way in which the grounds of the earth are indeed the very grounds of our planetary identities and that these grounds are being *blindly* wasted. In other words, this combats the foundational fantasy by choosing dirty grounds over pure foundations and by asking individuals to open their eyes to the reality that we are all extremely dependent on human and earth others. Second, it highlights the *mystery* of the ongoing process of nature naturing. The earth as grounds for identity and knowledge formation implies that transcendence becomes not something that is spatially beyond, but simply something that is radically not yet. We just don't know what the introduction of one solution will do five, ten, or fifteen years down the road because the earth and our own identities are always on the move. All is not decided, and the grounds of our identities and knowledge claims must rely on at least a little bit of agnosticism. Third, it places us in the mixed reality of how the environmental crisis may disclose common grounds. In other words, it took the very crisis we (Western and globalized Westerners) are in to begin to fully understand our own embeddedness within this evolving planetary community. All these things are necessary for forming a nomadic planetary ethic.[27] The only language I might challenge is the language of "beyond differences" that so many social justice and environmental thinkers long for, including myself at times. That smooth place beyond differences where we can finally land solidly and squarely in some type of promised land, or at least better place, is the imagined place that gets us in to trouble as we begin to try to build solid foundations in a reality of flux. Again, this type of language, which is hard not to long for, is powerful and it compels, but it may be problematic from this nomadic planetary perspective. As Keller and Kearns lay out in the introduction to *EcoSpirit*, grounds are not foundations and constructing grounds must always be *with* differences rather than *beyond* differences.[28] Too often this desire to move beyond differences in a world with highly polarized class/power differences leads to the backgrounding of difference in the name of a goal that is much more defined by those closer to the center of power/economic wealth.[29]

For these reasons, the argument for a nomadic understanding is not merely over academic epistemological disagreements with foundational understandings of nature. Where one falls on this matter may also be at heart an issue of ethics, more specifically one of postcolonial ethics. All empires or economic powers have centers toward which resources from the periphery are pulled. This central *place* literally exists at the peril of other places. Again, Brennan writes, "Spatial centralization creates energy demands and an energy field which can only sustain itself by extracting surplus value from nature."[30] In other words, the globalization of free-market capitalism is yet another form of colonization in that it draws more and more resources from the periphery in order to support the centralized industrial, consumer economy. The center is no tangible center, but an economic center supported by Bretton Woods–type institutions and multinational corporations and fueled by the increasing number of people who are forced (or buy) into the emerging global consumer society. This global movement of resources due to agriculture, technology, fashion, and durable goods is more recognized and adversely felt by those who live at the periphery. Those benefiting from the centralized networks that extract and move resources all over the world are essentially imposing their place, and their own ideas of place, onto the whole globe. Meanwhile species go extinct, forests are razed, oceans rise, and the climate is changing. This powerful metaphor of place and stasis is, as we have seen, dependent upon recreating the rest of the world at a pace that outstrips what Brennan refers to as generational time. This foundational, static understanding of place is indeed making much of the rest of the world barren/static. This is the cost of imposing transcendent values across the face of the globe and forcing all life to conform to those values.

Perhaps, then, an ecological nomadology (discussed further in the next section) may be more successful in highlighting the glocal nature of all places on the planet. In this sense it will address the issue of how peripheral places and identities are backgrounded in such a way that the center does not realize its own dependence upon them. In other words, perhaps a system of meaning making that took nomadism as its core trope might better capture the postcolonial and globalized experience of more peoples today than does one that takes as its tropes home, local, and place. After all, many religious traditions have deep roots in some form of nomadism: Aboriginal walkabouts, exile, pilgrimage, liminal journey, traveling disciples who are specifically told to leave behind their place and kinship identities to follow Christ. Though some of these examples involve arriving at a place, the journey is at least as important, if not more, in some cases. Instead of the arrival at a foundation, or a place, what type of ethic and meaning might be constructed from metaphors of traversing grounds and journey?

Again, in their introduction to *EcoSpirit*, Kearns and Keller posit the trope of common grounds for meaning, ethics, and knowledge over that of foundations. "To ground our thinking in the shifting and shared finitudes of present places enacts the *hope* of an intentionally common future through collective action and celebration."[31] Common grounds capture the changing and evolving realities of nature-cultures: ecosystems evolve, plate tectonics shift, earth and minerals erode and are blown about by wind and washed away by water, meanings are constructed and change over time, ecosocial contexts are always on the move. How then, can we make meaning in this ever flowing context? How can we construct ethical guides upon changing grounds? How might meaning-making processes converge on common grounds that are ever expanding their realms of moral concern to connect and include others, rather than enforcing transcendent forms onto places? Unlike finding the common ground above differences, this understanding of common grounds constructs hope through linking differences together in order to have a place to rest or dwell, if only for a cosmic moment. Rather than directing our attention away from the dirty differences that make up evolving planetary realities and identities, meaning-making practices become more akin to what Latour claims about the function of religion. He writes, "Religion . . . does everything to constantly re-direct attention by systematically breaking the will to go away, to ignore, to be indifferent, blaze, bored."[32] That is, religion rereads and reconnects ad infinitum in a way that is hopefully (but of course not always) constantly destabilizing and recollecting to include that which is excluded from any collection of meaning.

Thomas Tweed articulates a similar understanding of religion in his book, *Crossing and Dwelling*. He writes, "Religions are confluences of organic-cultural flows that intensify joy and confront suffering by drawing on human and suprahuman forces to make homes and cross boundaries."[33] He goes on to suggest that we need to map these flows in "spiritual cartographies," which can help us read how power and meaning are negotiated in certain times and places.[34] How are our meaning-making practices materializing in the world and who and what is left out or excluded? How are these meaning-making practices mattering bodies in the world and how do they help us to negotiate identities and common grounds in increasingly multiconnected worlds?

The shift in meaning from foundations to common grounds is a shift from vertical to horizontal meaning. The former is created in a no-place, or vacuum, that sucks all life toward it, much like the center sucks in the periphery. The latter is created within the space-time continuum, from ever changing contexts to no definite end. Horizontal meaning seeks connections between differences rather than the imposition of sameness over the face of the globe, moving toward what Gayatri Spivak identifies as a planetary reality rather than that of a globalizing one.[35] From a global perspective, meaning is conferred from a smooth place of foundations upon the rest of the planet. From a planetary perspective, differences are constitutive of any identity, and commonality takes place through negotiating these differences. This nontranscendent understanding of meaning and commonality is difficult to grasp for those of us who have been steeped in metaphors of transcendence, place, and foundations. Perhaps, and only a bit tongue-in-cheek, the contrast between the language of coping and flourishing might be helpful here. It will help us to discern the difference between a monotheistic and polytheistic nomadism.

From one perspective, and one I don't find promotes much joy or love (two experiences I highly value in life, constructed or not), the happiness or success of one's life can be seen as finding the coping mechanism that allows one to maintain a sense of order, self, and place against all of the existential anxiety, sadness, and messiness that is part and parcel of living in globalizing worlds today. This is not necessarily the Marxist "opiate of the masses," but does include the Marxist interpretation of religion as a coping strategy. The coping mechanism could be religion, reason, fantasy, or what have you, but the point is that you are strengthened in your own foundation when encountering the world rather than necessarily being open to challenges when encountering an other. You define yourself over and against others rather than in porous negotiations with others. This is what I refer to as a monotheistic nomadism, which amounts to foundationalism. In this type of nomadic movement, one carries one's god/meaning to all places on the globe. In the worst-case scenario, one enforces one's god/meaning over the face of the entire globe. From within this setting, an encounter with an other identity or place means denying the difference of that other in favor of the same image of one's god. This is akin to the contemporary problem of globalization: a single regime of truth is being forced over the entire planet at the expense of many identities, human and non.

Another perspective suggests that life is about flourishing. From this perspective, happiness is about throwing your life as far as it will go, so to speak. This perspective suggests that the process of making meaning is more important than necessarily identifying what is *ultimately* meaningful and what is not. Meaning is that which enables us to move along in life into the unknown not-yet future and hopefully toward the future flourishing of planetary identities. Meaning is still a regime or technology, but from this perspective there are many possible lines of flight toward the not-yet future that could lead to flourishing or atrophy. The point is not to identify *the* way, but *a* way, and to be open to changing one's way when it leads to planetary atrophy.

This mode of meaning making I would describe as polytheistic. From a polytheistic nomadic perspective, one is open to the contours and differences of human and earth others. One connects the world and places together into a planetary whole through differences rather than enforcing sameness on the entire planet. There is no Archimedian or vertical point from which one can enforce meaning; rather, meaning here becomes horizontal.

Again, sources for meaning making come from religious, philosophical, aesthetic, scientific, and many other maps, but these sources now become immanent to the planetary meaning-making process that we human earth creatures participate in rather than stand outside of. Thus, horizontal meaning-making practices provide us with various cartographies for understanding the difference in and between human identities and earth others, while at the same time acknowledging the ways in which these maps are not territory. These types of maps are dynamic rather than static. They are more like animated radar maps marking the shifts that flow into a given moment. As Heise suggests, we need to move from that image of the blue planet to that of Google Earth.[36] We need "a sense of planet" more than we need a "sense of place" in this planetary context where all of our local problems have planetary dimensions. This move would mark a change in our imagined communities from that of individual and nation to that of planetary citizens.

Horizontal meaning-making systems, such as a polytheistic nomadic ethic, may actually be more adept than vertical monotheistic ethical systems at opening humans onto the rest of the natural world to include the nonhuman world in the coconstruction of common grounds and assemblages rather than keeping us locked into vertical transcendent foundations of isolated individuals and strict species boundaries, for example. They do not lock us into discrete identities of what it means to be human/other, other/self, or female/male, but rather take note of the many different confluences and flows that make up a given dwelling or assemblage at any given moment. These horizontal understandings of meaning do provide us then with places to dwell. As Tweed reminds us, "Dwelling is always 'for a time,' it is never permanent or complete."[37] We dwell then, on horizontal terrains moving toward uncharted territories.

The Anthropological/Political Level: Uncharted Territories

Žižek writes, "Indeed, what we need is an *ecology without nature*: the ultimate obstacle to protecting nature is the very notion of nature we rely on."[38] Žižek, Morton, Latour, and a growing number of other poststructural thinkers see a foundational understanding of nature as one that has been used to justify oppressions and the status quo through claims to what is natural or unnatural. One need only think of the recent findings of Ardi and the subsequent rush by primatologists and anthropologists to justify heterosexual monogamy (Lovejoy) or the peaceful origin of *Homo sapiens* (de Waal).[39] Or, less recent, the rush to find the gay gene in an effort to make homosexuality normal or natural. And of course there is the constant effort to justify superiority of race and sex through whatever science is contemporary, as was discussed in the previous chapter. Finally, there are the colonizing effects of the Western understanding of nature as not including humans (which will be discussed further in the next chapter): from the early conservation and national park movements that removed indigenous peoples from their lands to contemporary struggles between international environmental groups and indigenous peoples over hunting and fishing of various animals.[40] The list goes on.

A polytheistic nomadic understanding of environmental ethics would not have such a stable concept of nature as foundation to draw upon and would thus problematize essentialist identity politics. Again, there is a way in which place-based ethics mimic the essentialisms that take place in identity politics. To be woman is to be X. To be homosexual is to be X. To be identified as environmentally sound or respectful is to be X. To be natural is to be X. When we get down to it, these identities are all messy assemblages, and there is no essentialist way to seal the boundary between one and the other. This, again, is a manifestation of the foundational fantasy in the form of the Lockean liberal self. The Lockean understanding of self (directly tied to production and private property) represents the epitome of the individual self. Much like the omni-God needs no other to create, so the Lockean self need only mix his labor with dead matter, or have other labor and technology do that for him, in order to create property (capital).[41] The more this liberal self moves away from his own labor and toward the use of capital, technology, and labor of others, the more he is reconfirmed in his omnipotent power.[42] The collection of selves that then find themselves at the center of power define what is natural and have the economic backing to enforce those definitions across the face of the globe. Of course, the liberal self must, as Plumwood argues, background its actual dependence upon human and earth others (through creating "shadow places")—and at a very high cost for those others. It is this type of liberal self, and the cultural memory of that self, that may be at the heart of the drive for arrival at place. And this desire and construction of place may have some of the same violent consequences for human and earth others as the desire for and construction of the liberal self.

Perhaps just as the foundational fantasy of discreet subjecthood mistakes self as Self, the place-based approach of connecting with nature mistakes its own understanding of nature with Nature. In fact, place-based approaches do emerge from certain social, economic, and even racial locations. Some place-based approaches may even smuggle in a type of heteronormative family values. To be faithful to a place means to be in a monogamous relationship with a single patch of earth: to stay put, forever wed. Wendell Berry's place-based approach has also been critiqued for favoring a patriarchal family structure, for instance, and thus it also promotes heteronormative arrangements as normative.[43] Monogamous lifetime relationships and the longing for them can be a part of the utopian dream to return to a paradisiacal past. Just as globalization brings together distant places in ever flowing relationships, so it brings together distant peoples into relationships. These relationships are not just external meetings of subjects and objects; rather, each one changes the makeup of one's own identity. Our conscious identities are literally the results of the interactions we have with human and earth others over a lifetime: each of these external experiences changes our internal neuronal structures.[44] Identities are always already nomadic, unfinished, and on the move.

The desire for the monogamous romantic pairing—a historically recent norm for marriage and coupling—may not map onto the nomadic, evolving structure of our planet today. In a time when communication and transportation enables mobile identities and in a time where economic forced migrations tear families apart, we simply have more opportunities for multiple connections: we have shifted from dial-up place-based technologies to those of digital, wireless, and nonplace-based ones. We no longer (especially those of us in industrial and postindustrial nations) live our lives in small isolated communities where it makes more sense to choose/settle on one person. Yet, just as we long for place, many (if not most of us) still long for that one person. Does this longing not perform an apocalypse on the value of past relationships and possible future relationships, just as the longing for place never enables us to be in other places fully? In fact, the very economics associated with a monogamous relationship to a place, also known as having the monetary resources to create the illusion of maintaining a single place, may exacerbate environmental problems. Just as maintaining the notion of a consistent self often occurs at the expense of many others, so fidelity to a specific place may be at the expense of the many relationships and dynamics that affect a given place. In other words, the postcolonial insight that centers of power exist and persist because they draw resources and energy from the hinterlands to secure the privileged place of the center is apropos here. Note, I am not suggesting that monogamy and place are always *wrong*, but rather they have been privileged to such an extent that much energy and resources are used to maintain them, even if it means the deterioration of multiple and many earth others.

Perhaps, then, a polytheistic nomadic understanding of identity would foster a little polyamory or love relationships with multiple places. Maybe a little love on the run would help us to make the type of planetary connections necessary for an ethic that resists forces of social domination and ecological ills. After all, multiple places already flow into identity formations, and this type of nomadic identity may be what characterizes anthropology for the foreseeable future. This type of identity formation moves us into uncharted territories.

The more culturally, intellectually, environmentally, politically, and economically globalized our worlds become, the less effective will be place-based politics, whether that place is a state, a bioregion, a township, a neighborhood, or a nation. These once seemingly rigid boundaries around places due to less advanced communication and transportation technologies are now revealed as permeable and often arbitrary. Our old maps for addressing problems do not necessarily work for this new context. How, then, do we develop an adequate environmental ethic on the planetary scale? How do we develop a political movement that works outside the defunct boundaries of local/global, nation states, self/other, and near/far in order to address the already global capitalist assemblages that negotiate, move, and work regardless of one's place? Perhaps a place-based approach is merely an acceptable mode of environmental protest precisely because it does not challenge the current economic hegemony, which is a place-less center that is transnational, globalizing, and thus able to easily resist isolated pockets of protest—much in the same way the narrowing of the LGBTQ movement to gay marriage is an acceptable form of protest because it formulates love in the monogamous, heterofamiliar form; despite protests, it is probably the most acceptable, or economically and politically manageable, form of LGBTQ relationships.

I think a trans context that challenges the normative centers of place and self is already emerging with the work of organizations such as the Forum on Religion and Ecology, the United Nations Environmental Program, and the Earth Charter, just to name a very few. Still, we need ever expanding creative imaginings, or what Deleuze and Guattari call lines of flight toward the unknown, open future. We need ever newly imagined ways of becoming that persuade us to develop technologies and live in ways that respect the (re)generative energies of the planetary beings we are becoming. We need truths we can live toward rather than ones that tell us who we are and what we must do. As Latour notes, "Truth is not to be found in correspondence . . . but in taking up again the task of *continuing* the flow, of elongating the cascade of mediations one step further."[45] From within this nomadic polyamorous understanding, meaning-making practices are more about the art of persuading us toward a vision of a future that is uncharted. A future that is not-yet (*à la* Ernst Bloch) is a future that could matter in many different ways.

Components of a Planetary Environmental Ethic

What might be some suggested actions from within a planet that is on the move? What does it mean to act when our realities and even our identities are constantly in flux? Again, as argued throughout this book, I think this has always been the context, but how do we both acknowledge this context and take responsibility for our actions in the world? Surely an ethics that makes universal claims or one that asserts acontextual principles will not be the best response to a planet on the move. Rather, we must begin to think of ethics in terms of movement. We must begin to create the space to think about the ways in which our movements affect others and the rest of the natural world. In a sense, contemporary ethics based on universal claims and principals are just a shortcut for this type of thought, and rightly so since the processes of globalization seem to increase our experience of time exponentially so that we literally don't have time to think. Rather than write such messy, time-consuming projects off with a "who has time for that, we need answers" mentality, I would argue that it is precisely books such as this and courses in religious studies and the humanities in general that should give us the space-time to begin to think about an ethics of movement. As such, I offer here not solutions to our most pressing environmental problems, but suggestions for what might open up the space-time for us to begin imagining ourselves as planetary beings. First, let me briefly review the contemporary context in which many readers might find themselves.

In contemporary U.S. society, students exit college with a debilitating amount of student loan (and other) debt. Furthermore, health care is tied even still to employment, though the Obama administration is working hard to change this. Couple this with the dismal job market, and you have a whole host of people in the late Gen X and Millennial generation who have a lot of debt, no health care, and no viable job prospects—not to mention baby boomers who will likely live long past the ability of Social Security and other social nets to handle. This leads to existential panic, which overwhelmingly persuades one toward finding any job just to make ends meet. Further, it fuels the drive to work longer and harder, thereby leading to a labor force that is unhealthy and has little time to think anew. It is a labor force that will outstrip the regenerative capacities not only of the planet but also of human imagination. If the recent Occupy movement is not in response to these issues, then I'm not sure what it is all about! Where is the space-time for new generations to think together imaginatively? The background conditions are stacked against this type of creative/imaginative/existential thought. Even as new environmental and social problems arise, the most this generation might have time for is to read the barrage of information on the Internet and click on a button to donate money to some organization (a form of "slacktivism," as I once hear it called). Boomers and older generations complain about younger generations wanting instant gratification, but those older generations have helped to contribute to the conditions that create the system younger generations are born into. In a world that moves faster than there is time to generate a thought, all one has time for is instantaneous gratification, quick thinking, and action. One must move on or be left behind. The God of globalization indeed saves those elites in the top 20 percent of income and resource use on a global scale, while the rest (including the rest of the natural world) get left behind, suggesting a terrestrial apocalypse of sorts.

In light of this somewhat overdramatized scenario, what might be some possible steps toward creating the space-time to become planetary citizens? How might religions contribute to a planetary nomadic, environmental ethic, an ethic that both acknowledges the reality of planetary movement, but also has the critical space to examine the speed of movement? The intersection of religion and ecology can, I argue, provide some of the critical tools needed to open up such spaces, but thinking about the causes of environmental crises and our responses to them must be expanded beyond those of mere efficient causality.

In 2004 Michael Shellenberger and Ted Nordhaus shot a flare across the bow of environmentalism with their policy paper "The Death of Environmentalism." Though I don't agree with the tone of the paper, nor with some of the facts, what I do agree with is the general sentiment: what the environmental movement has been doing is not working anymore, and we have to step back and take a look at what kinds of things our energies should be spent on in order to make some real changes toward a more ecologically sound and socially just future. This, for instance, has also been at the heart of the Environmental Justice movement, which they totally leave out in their paper.[46] Again, the basic call to rethink environmentalism is, I think, right on.

In a globalized world that is more and more marked by what Zygmunt Bauman, among others, refers to as the space-time crunch, background conditions matter more and more. In other words, saving the rainforest, or saving the whales, or cleaning up rivers and streams, need more than just direct actions aimed at completing those tasks. Mind you, those are still important and necessary, but not sufficient. This issue-based-solutions approach to ethics is a feature of the shrinking of causality to mere direct/efficient causality. As Terrence Deacon's work on evolutionary emergence has argued, we need to remember the other forms of causality as well: material, formal, and final.[47] Much of the field of environmental ethics, including religious ethics, has been focused on efficient (praxis) causes and final (aims/purposes) ends. I am not arguing that we need to stop focusing on these types of causes, but rather, we ought also to focus on the material and formal causes.

Let me be clear, I am not suggesting that this work has not begun. The work on religion, ecology, and economics by scholars such as John Cobb, for instance, are addressing some of these formal causes. The work of ecofeminists deconstructing patriarchy addresses some of these formal causes as well. However, the transition from identifying these formal causes often slips into solutions that fall in the trap of attaining final ends/goals through efficiency. For example, the problem of global climate change. The solution: respecting God's creation and finding our proper place therein (final) by recycling, driving less, eating lower on the food chain (efficient). Or, consider the problem of economic inequities that lead to greater environmental ills for poor people, especially poor people of color. The solution: recognizing that ecological and social degradation go hand in hand and working toward a more ecologically just future (final) by supporting women's microbusinesses in third world countries, eating local, organic, fair-trade foods, and writing representatives about unjust trade policies (efficient).

Again, the problem is not that these things are superfluous and useless. They are necessary to, be sure, but they are not sufficient. Identifying material and formal causes, in addition to efficient and final causes, of our contemporary planetary ills may then lead to some complementary solutions. Here I will focus on three of the formal causes of planetary ills. They do not seem like environmental issues at first, but these formal causes are at the heart of whether or not there is sufficient creative space-time for thinking toward the future of planetary realities. In other words, these formal causes are often roadblocks between efficient causes and final ends. Through examining them, hopefully new spaces for planetary political thought will begin to emerge.

Free Higher Education: Bailout for Students Crushed Under the Weight of Student Loan Debt

As undergraduate and graduate students in religious studies, philosophy, and the humanities especially, we are often trained to identify efficient causes (the problem) and work toward final ends (solutions to these problems that, ideally, work toward a better world). However, the formal causes frustrate this leap once the realities of the job market and economics of education set in. Student loan debt is crushing the space-time that allows for future generations to contribute new, creative solutions for planetary problems. At the end of one's education, especially if one's passion falls within the humanities and the arts, one is faced with fewer and fewer jobs and more and more debt. This pressure cooker can lead one toward despair and self-loathing for not being able to work toward those final causes; eventually it can lead to settling for some type of work just to make ends meet. In other words, these conditions help force us into the coping strategies of foundational thinking mentioned earlier in this chapter.

Might ending the crushing debt of education help create a habitat for thinking toward a better planetary future? Even further, given that the economics of education create formal conditions that pressure people into professional vocations such as medicine and business, which will eventually pay off those student loans, might free education lead to the graduating of more students with humanities degrees? Not that an MBA and an MD are thoughtless degrees! They are degrees that gear thought toward specific, professionally defined outcomes rather than degrees that gear thought toward critically questioning the formal definitions and structures set forth by economics and health, respectively. From this perspective, ending student loan debt becomes one of the top priorities in the one-fifths world for imagining a viable planetary future. Most students now graduating, at least the privileged ones, are steeped in the problems of global climate change, overconsumption, overpopulation, bleaching of coral reefs, and species extinction, but they have not been given the space-time habitat to address these problems with new, creative ideas. The formal structures that give rise to these problems disallow the imaginative space necessary to think through a transformation toward a radically new way of thinking about what it means to be a human in a planetary context, and so suggested solutions to the problems are piecemeal at best, never addressing the underlying structures that maintain and support these destructive ways of being in the world.

Universal Health Care

For many of the same reasons that we need to relieve student loan debt, we also need universal health care. Our greatest connection to the rest of the natural world is through our bodies, yet many graduating students are left without the means to care for those bodies. This creates great anxiety and a narrowing of focus from "How am I going to address these social-ecological problems?" to "What am I going to do when I get sick or I need a root canal?" Again, these are privileged questions from the privileged context of the West. But, that is the context from which I write. What I am getting at here is the question "What can privileged Westerners do to contribute to a better planetary future?" Oftentimes, we sidestep our own context and go for obvious efficient solutions without even thinking about the formal causes that perpetuate planetary problems in the first place. Universal health care not tied to work or marriage would do a lot to create that space-time necessary for creative and constructive solutions.

Promoting More Leisure, Play, and Meditative Space-Time

Religious communities are particularly poised to push for more leisure and meditative space, and the religion and ecology/nature nexus is particularly well positioned to influence this space toward planetary reflections.[48] Part of the problem, again, is the space-time crunch. There simply isn't enough time to bring the past and future into a present moment of conscious creative transformation. Here, I am not thinking of only promoting traditional forms of meditation, prayer, and retreat. These things are good, but again, not sufficient. Another role that religious communities could play (and some do) is just creating habitat for fun, community, and reflection regardless of whether or not it has anything to do with what we might consider to be religious. Fat Tuesday celebrations, informal conversations on current events, opportunities for play such as trips that take people water and snow skiing, fishing, to the beach, or discussing theology and philosophy over beers are some of the free spaces that I am suggesting we need more of here. The Millennial students that fill classrooms now are bombarded with information from day one. They don't need more indoctrination; they need space to breathe, to play, and to reflect beyond the boundaries that are imposed upon them by the structures they are born into. As Juliet Solomon argues, "By the time we are adults, we have ceased to allow ourselves to indulge in spontaneous creative expression in physical movement" and thought.[49]

These are just three small examples of what I mean by an ethic of movement, which challenges a narrowing of causality to that of efficient causality. An ethic of movement must look at formal and material causes because it deals with the flows of information. Flows are guided by formal causes such as ideas, habits, and institutions, and they are made of material causes, namely, matter and energy. Efficient actions will fall short without addressing these formal causes, and final ends are impossible to imagine if there is not the free space of critical imagination necessary to chart a path toward the future. There are many other examples, such as tax reform that would end taxes for those making less than X amount per year, not merely reform of environmental tax laws; reforming marriage so that many different types of unions would be recognized, again, alleviating the economic and cultural incentive/pressure to get married in the first place; and, something that has been proposed, providing high school seniors with more options to do things such as AmeriCorps or Peace Corps that provide space for engaging the world and "soul searching."

These practices may help foster a planetary environmentalism and spirituality, for lack of a better term. Such a planetary outlook allows us to travel through many places, receptively open to the contours of those places and peoples, which is at the heart of a polytheistic nomadism. It has the capacity to enable deep attention to and appreciation for the many human and earth others we come in contact with on a daily basis: the crux of a polyamory of place. At the same time, such an ethic of movement through places may shed light on how attention to only efficient and final causes background the interconnections that go into any given construction of place. In other words, it is not enough to focus on final goals and address the efficient causes that are constructed and constrained by the very formal causes we need to change; rather, the formal causes of the system are out of whack with the material causes, and these formal conditions must be addressed if we are to live toward a more ecologically sound and socially just planetary future. Such a planetary ethic recognizes that we are always already nomads. We are always already involved in political projects of persuasive truths. Perhaps we can find a means to live within this nomadic, evolving process in a way that does not seek to reify the present, but rather allows us to dwell in ever new realizations along with the rest of the evolving planetary community. As we open onto the rest of the planetary community, we open ourselves to becoming something other, which is the topic of the next and final chapter of this book.

7 CHALLENGING HUMAN EXCEPTIONALISM

Human Becoming, Technology, Earth Others, and Planetary Identities

Perhaps intentionality might better be understood as attributable to a complex network of human and nonhuman agents, including historically specific sets of material conditions that exceed the traditional nothion of the individual.
KAREN BARAD, Meeting the Universe Halfway

There was never a time when human agency was anything other than an interfolding network of humanity and nonhumanity; today this mingling becomes hareder to ignore.
JANE BENNETT, Vibrant Matter

Here, in the final chapter of this book, we encounter a boundary writ ultimate by religious and scientific thinking about nature and truth: that of our very species. To a great extent, our species, like our subjective experiences, is always already not our own. In order to claim a species boundary one must make out of the present evolutionary moment an ultimate foundation and background the millions of other species, plants, animals, minerals, and other organisms that have produced this very moment. Such space-time belongs to the linear space-time of eternal foundations, linear progression, and monological identities critiqued throughout this book. What we need is a new way to think about our species as a nomadic organism moving toward an open future, just as we can think of ourselves as always already a part of multiple earth others on the move.

Time, for the nomad, is not about eternal return or linear progress, but more about infinite recapitulations, infinite per/versions, and infinite journey. The present holds past and future together, but no spot is privileged: all are here together. What becomes less important is a past *in situ*; that is, it matters more how the past is narrated and taken up by the present. Likewise there is no present *in situ*, only gathering up, reconstructing, growing edges, and horizons that are constantly in movement, shaped by past and future. Finally, there is no future *in situ*, but future becomes the event that enables the continuing movement of natura naturans. There is no end, only more diffracted possibilities toward which life moves. As Žižek notes, "'Eternity' is not atemporal in the simple sense of persisting *beyond* time; it is, rather, the name for the Event or Cut that sustains, opens up, the dimension of temporality as the series/succession of failed attempts to grasp it."[1]

From this position of radical immanence, space and time are interconnected, as Einstein imagined. Movement in space is the time of planetary and cosmic reality. Space is nothing but the gathering and recognition of movements from particular, situated moments in time. This is not a valorization of movement and newness. Indeed, as Brennan suggested, and as discussed in previous chapters, we have moved into a period where our own technological capacities have outstripped planetary time. This does not mean that technology and humans are somehow unnatural. Rather, it means that humans, as meaning-making creatures, have most often been narrowly interested in the pace of human time.

Humanity is but one among many fuzzy conglomerations in space-time, yet we try to recapitulate all reality into our own time. The task then may be to try to understand better the space-time required for planetary flourishing. This is one implication of Deleuze and Guattari's discussion of becoming plant, animal, and mineral. In speaking of contemplating animals, plants, and minerals, they write, "These are not Ideas that we contemplate through concepts but the elements of matter that we contemplate through sensation. The plant contemplates by contracting the elements from which it originates—light, carbon, and the salts—and it fills itself with colors and odors that in each case qualify its variety, its composition: it is sensation in itself."[2] It is the same sensorial way in which the human subject works. Subjectivity is nothing more than the assembled ongoing experience of different things giving rise to contemplation. Plants, minerals, animals, and humans are all various contemplations on the continuous flow of experience. In order to open up further possibilities for the future, human beings can enter hybrid relationships with other life in hopes of creating lines of flight for future possible becomings. "The human becomes more than itself, or expands to its highest power, not by affirming its humanity, nor by returning to animal state, but by becoming-hybrid with what is not itself. This creates 'lines of flight'; from life itself we imagine all the becomings of life, using the human power of imagination to overcome the human."[3] Perhaps a recognition of becoming *with* the rest of the natural world, including technology, might open us onto a planetary time that does not outstrip the earth's generative capacities.[4] Furthermore, it opens us up to different ways of becoming human, and possibly posthuman. Such a method or technology of understanding is different from the "global ethic" called for by thinkers such as Hans Kung. Kung, though venturing far into intercultural and interreligious dialogue in such a way as to make this writing possible, doesn't do much to move beyond the human model. The model being outlined here must move beyond the merely human and beyond thinking about the "core commonalities" that are needed for planetary dialogue. For reasons suggested throughout this text, what we need is to join together through our differences rather than our commonalities, as the race to the common often only supports mere toleration of difference in the name of such commonalities.[5] Though statements such as the Universal Declaration of Human Rights may be helpful in theory, in reality such rights are not secured by these common declarations. Furthermore, a rights-based discourse does not meet the messiness of subjectivities that are always and already multiple, messy, and evolving and implicated with the rest of the natural world.

New assemblages emerge in nature, but not from top-down ordering of global impositions; rather, there must be experimental linkages of differences from the bottom up. To what degree these assemblages are biological, evolutionarily genetic, technological, or virtual/imagined is a faulty line of questioning; such conjecture is still within the hangover from a metaphysic of substance and an epistemology of representationalism. All these aspects are always already involved in nature naturing. The not-yet space of emergent newness is just as much a reality for the rest of the natural world as it is for humans. The difference in humans, as meaning-making creatures, is that in our consciousness awareness of the not-yet and multiple possibilities for moving toward the not-yet emerge. In other words, response-ableness becomes responsible. A better line of questioning might examine how emergent assemblages return to affect the past, present, and future of planetary becoming. Who/what will flourish under these assemblages and who/what will not?

Here in this final chapter, I will flesh out some of these material ideas in an attempt to think with the imaginative capacities of the rest of the planetary community. Such imaginings are required if we are to move out of our speciesist ways of thinking and create planetary communities. As discussed in earlier chapters, if truth regimes are like lures that persuade us into habits of thinking and becoming, here I offer one lure out of our comfortable habits of being human. I realize that some peoples are lured by technology and urban landscapes, and some are lured by mountains, rivers, forests, and streams. I argue that it will take the lure of organic machines to help us finally work beyond the confines of humanity and open up onto other possibilities for planetary becoming. The skyline of San Francisco, the Muir Woods, the sun rising over the Atlantic Ocean, the moon rising over the skyline of Miami, the Everglades and Half Dome at Yosemite, the enchanting water towers on tops of the buildings in New York, the Hindu temples of Indonesia, the vineyards of France, Spain, and California: all these things possess beauty and all are equally part of nature naturing, along with so many other not yet emergent possibilities. What might it mean then, to become plant, mineral, and animal? What might it mean to become cyborg? Proffering persuasive answers to these questions will in the end mean talking about what it means to be but parts of hyperobjects/hypersubjects becoming toward a planetary unknown.

Becoming Plant, Mineral, Animal

> The error we must guard against is to believe that there is a kind of logical order to this string, these crossings or transformations. It is already going too far to postulate an order descending from the animal to the vegetable, then to molecules, to particles. Each multiplicity is symbiotic; its becoming ties together animals, plants, microorganisms, mad particles, a whole galaxy.
> DELEUZE AND GUATTARI, A Thousand Plateaus

From one perspective, this whole book is about moving from definitions based on boundaries, certainty, and borders to looking at our categories and identities as ecotones, *mestizaje*, fluid, and permeable. As Barad notes, this involves a process that resituates knowledge from something that is representational to understanding knowledge as performative. In order to make such a shift, it is not enough to talk about the performativity of human beings, but rather we also have to talk about the rest of the natural world. In other words, the tools of queer theory, poststructuralism, deconstruction, and other critical theory need to be expanded beyond the human species boundary to include the rest of the becoming planetary community. Many have begun to do this type of work, and this final chapter brings some of these voices together in an effort to think beyond the boundaries of human exceptionalism.

Many may think to themselves, "of course I know that humans are animals and that other animals ought to be treated as subjects." However, human exceptionalism is ingrained in the humanities, and especially in the liberal understanding of the individual that determines our economics, politics, laws, and social and individual realities.[6] The problem is that such humanism is inherently flawed. All one has to do is think of the illogical reduction of all animals to either use or nonuse.[7] In the United States, for instance, cats, dogs, parrots, and other animals serve as pets, while cows, chickens, pigs, and other animals are seen only as resources. These pets come to have the same rights as individual humans while the economic animals have zero rights, similar to human slaves or certain groups of people who have been subject to genocide. As Plumwood and others have pointed out, such arbitrary distinctions between use/nonuse are problematized by cross-cultural analysis. Furthermore, such use/nonuse distinctions cover over or background the fact that all of life is part of a predator-prey process and to a certain extent all life is marked as use/nonuse, even human life.[8] Human exceptionalism is that rule by which we do not understand ourselves as part of the predator-prey cycle: even in death we seek to keep ourselves from returning to dust. Or, to put it in another and perhaps more *verboten* light, we might critically analyze the taboo of bestiality. Why is it that most people are horrified at the thought of sexual relations with other animals, yet when it comes to brutalizing them in factory farm–style settings and eating their muscles, blood vessels, and body parts the same people do not blink an eye? I'm not advocating for bestiality, nor am I advocating for what Plumwood calls "ontological veganism."[9] Ontological veganism, according to Plumwood, simply extends the flawed category of the Lockean individual to other nonhuman animals and thereby seeks to take them out of the ongoing predator-prey cycle. What we need, rather, is to rethink our own selves as part of the ongoing planetary process of creative destruction. One way to do that is by thinking seriously about ourselves as becoming plant, animal, and mineral.

Deleuze and Guattari challenge the very notion of the boundaries of subjectivity and humanities in their discussions of becoming plant, animal, and mineral. It is not just that we are made *of* histories and biologies of evolving plants, animals, and minerals, nor that we will become part of future plants, animals, and minerals. Rather, it is that these companions literally make up our multiple, evolving, and open subjectivities. Just as queer theory recognizes our subjectivities as always already multiple, so from a radical materialist perspective we can say that our embodiments are always already multiple. As such, our agency is not just the agency of the Cartesian skin-encapsulated ego, nor are our thoughts and emotions our own. Our actions, thoughts, and emotions are always multiple. They involve multiple histories of planetary becomings or communities of plants, animals, and minerals, all of which are evolving beyond their own boundaries and diffracting into proliferations of subject-objects.

There is no inherent reason why we ought to think of the boundaries and categories that we cocreate on a human scale as representative of reality. Instead, we can think of performativity "all the way down."[10] Like Deleuze and Guattari's understanding of becoming plant, animal, and mineral, Haraway begins to think with the rest of the becoming planetary community through her concept of companion species and Barad through her discussion of performativity. It is to a discussion of each of these thinkers that I now turn.

For feminist philosopher of science Donna Haraway, human beings have been evolved *by* just as much as they have altered the evolution *of* planetary others. In other words, our biohistories or nature-cultures go all the way down. History is not just for humanity, and maintaining that distinction is yet another divide proffered by anthropocentric ways of thinking. Haraway argues that we miss a lot if "the fleshly historical reality of face-to-face, body-to-body subject making across species is denied or forgotten" when "the humanist doctrine that holds only humans to be true subjects with real histories" is maintained.[11] We can talk about history in terms of our planetary others and in terms of biology; conversely, we can talk about the biology of history. This position is one that Haraway wages between biological determinism and complete constructivism, both of which she understands as falling into a form of reductionism. It is not enough for Haraway to understand history as an emergent feature of *Homo sapiens* alone. This would be yet another way of securing the species boundary that leads to human exceptionalism. Rather, she might argue along similar lines, as do some emergence theorists, that even single cells contain history. Again, the ability to respond, even if ins a fight, flight, or reproduce (the third f) response, already implies some sort of rudimentary recording of events. How else would organisms learn? We may prefer to call such rudimentary understandings of response-ableness "instinct," but we ought rather to think of response-ableness along a continuum from instinct to self-reflexive decision making. Anywhere one falls along the continuum always already implies biology and history, nature and culture.

Denying the rest of the natural world history is most often done for arbitrary reasons. Most often, such denial is done for fear of the charge of anthropomorphism. As Plumwood notes, the charge of anthropomorphism is one that is usually made just to reinforce the species boundary. In other words, the sign language of a chimp and the sign language of a human are treated very differently. Many suggest the chimp's, parrot's, or another animal's tool or language use is merely instinct or mimicry, while a human's use of tools and language represents higher-level cognition. We ought, rather, to use a modified form of the Turing Test, which suggests that if it looks, sounds, and performs like history, we should call it history.[12] For Plumwood, there will always be some amount of anthropomorphism precisely because we are humans describing and understanding the world from our human subjectivities, but this is no reason to deny that we may share some qualities with other animals and other planetary creatures.[13] Such a move would suggest that, just because I am trying to understand you from my own subject position, I could never say anything about you because I would always be imposing my own thought upon your subjectivity. Of course, we always already do this, but this is no grounds for not communicating as if we were reaching some common ground. Furthermore, our agencies and subjectivities, according to poststructural and queer theories, among others, are always already not our own but multiple.

The concept of companion species, for Haraway, is about the recognition that our subjectivities, identities, and agencies are not our own. We are multiplicity through and through. How could I type these pages without the food that gives my body the energy to do so, without the evolutionary adaptations that led to the formation of an opposable thumb, without the air that I breathe and the sun that plants use in the process of photosynthesis to create oxygen? The thoughts that appear on these pages are in dialogue with centuries of other thinkers who are all, in part, voices that shape my own most intimate thoughts. The animal familiars, be they pets or even those that I have come to admire through *National Geographic* or other nature shows, that are a part of my evolutionary history and my very way of thinking about what it means to be human are also spilling forth through the words I have written that appear on these pages. The isolationism of human exceptionalism narrows all these words to me and makes an ontological-epistemological cut at the boundaries of my skin that are just not possible without removing me in some sort of vacuum from the rest of the planetary community—an experiment that would result in my certain and sudden death. To begin to think and act with these planetary others, we have to begin to perform differently with our planetary others on a day-to-day basis. Barad offers up a helpful notion of performativity that extends beyond the confines of gender/sexuality and humanity to the rest of the universe, all the way down to the quantum level.

Barad, following an in-depth discussion of the quantum physics-philosophy of Niels Bohr, argues that "matter is produced and productive, generated and generative. Matter is agentive, not a fixed essence or property of things. Mattering is differentiating, and which differences come to matter, matter in the iterative production of different differences."[14] In other words, and much like the emergence theories discussed in earlier chapters, matter is not a dead collection of objects that human agency or consciousness is added to, but it is agential all the way down. Throughout her text, Barad argues that the epistemology of representationalism is that which makes a cut between epistemology and ontology, humans and the rest of the natural world, value/meaning and objective matter. Such a cut goes against the quantum indeterminacy argued for by Bohr. Even Heisenberg's uncertainty allows us to stay within a representationalist model of reality whereby our thoughts/values/meanings are somehow separated from the rest of the natural world and it is just a matter of achieving more or less accurate representations of reality.[15] Instead, indeterminacy means that epistemology is "ontoepistemology" and that ethics, epistemology, and ontology are always already together: matter is agential all the way down. There is no original or right way for reality to be. Nor are material things or objects the basis of reality. Rather, phenomena arranged by apparatuses are what order our reality into the things we see. "Apparatuses are not mere observing instruments but boundary-drawing practices—specific material re(configurings) of the world—which come to matter."[16] We human beings and each of our subjectivities are also phenomena that have come to matter. They are specific trajectories or habits of nature naturing. It is not that our modern scientific technologies are representing the world in a more genuine way today than, say, the technologies of ancient Egyptians or the technologies of other species or even a single cell. Rather, we are produced by and reproduce apparatuses that write humans out of the rest of the natural world through modern Western science. We are literally iterating and being iterated by the becoming truth regimes of this way of organizing the boundaries of various becoming phenomena. In a major way, Barad extends the notion of performativity to the quantum level and suggests that performativity ought not to be limited to human sexuality and gender studies. Rather, it also queers the boundaries of human exceptionalism by (re)placing us in a wider, planetary, and quantum reality in which performativity, truth regimes, and becoming phenomena subvert subject/object, inside/outside, observer/observed realities.

This is not the first time that queer theory has been extended to think about the rest of the planetary community. In particular, the volume by Daniel Spencer, *Gay and Gaia*, highlights an earlier attempt in the context of religion and ecology.[17] The radical fairies and lesbian back-to-the-land movements also draw out connections between queer theory and environmentalism (or thinking beyond human exceptionalism). However, much of this earlier bringing together of queer and environmental thinking (without which I would not be writing this!) also falls prey to the foundational or pure understandings of nature that tend to leave out the urban, technological, and historical side of the spectrum. In other words, technology, culture, and history are also part of the evolving planetary community. This side of the equation is also part of what is evolving humans beyond their species boundaries and other plants, minerals, and animals beyond their identities. In fact it is this technological side of things that likely allows queer and non-normative identities to persist and exist in relatively open ways in many places around the planet today. Hence, we must also talk about becoming cyborg.

WE ARE HYBRID, CYBORG, BIOHISTORICAL

One result of the apparatus by which human exceptionalism reproduces itself is that it reads technology or *techne* as a mere instrument for use by human beings to recreate the dead material world. If culture, history, and technology are apart from the rest of the natural world, understood as unnatural or human made, then all of that which is recreated through technology becomes part of the apparatus of human exceptionalism. How many of us consider the city to be part of the rest of the natural world rather than a fortress that keeps us out of nature? This is the reason that the mere romantic return to nature championed by many environmentalists does nothing to upset human exceptionalism. In order for this return to occur, must humans give up the very meaning-making processes and mechanisms that help make humans human (as we know ourselves today)? Does the fact that this version of environmentalism writes humans as somehow fallen away from nature not in itself reinforce the very notion that humans have become exceptional to the rest of the natural world? Just as Latour argues,"we have never been modern," so here I argue we have never been exceptional.[18]

In order to move past human exceptionalism and the anemic humanism that keeps making out of the arbitrary boundaries of the human species an ontological gap, thinkers such as Haraway have proposed a cyborg ontology. Such an ontology makes the boundaries between humans/animals, organism/machine, and material/energy permeable and soft. In other words, it suggests more of a performance of these boundaries rather than any sort of ontological/

metaphysical necessity for such boundaries. Furthermore, we are *created by* as much as we *create* these various technologies.

Part of moving beyond human exceptionalism and the subsequent anemic humanism that sees the rest of the world as standing reserve for use by human beings is understanding that we are not in complete control.[19] Biohistorical systems, which are larger than any given manifestation of an individual and cocreate that individuality along with the open possibilities of future becoming, also contain agency. Let's take, for example, the multiple versions of the technology of language.

Some languages will have multiple words or descriptions for what other languages see as a single phenomenon. For example, the Inuit of the Arctic region have about ten words/descriptions for snow, whereas in English we tend to use *snow* (with the variations of powder, wet, and packed). The Greeks have three words for *love*, whereas in English we tend to use one to describe sisterly/brotherly, erotic, and divine love. Being born into a biohistory in which this language shapes the way you see reality helps you to cocreate realities in different ways and shapes your possibilities for becoming. For another example, think about how rarely the passive mood is used in English as opposed to many other languages. Rather than the emphasis being placed on the occurrence or action in a sentence (passive: "truth be told"), the focus is placed on the individual agency of the person or thing doing the action (active: "tell me the truth!"). In fact, most students probably live in fear of using a passive voice, as we have been trained not to. So it is that even in our various technologies of language we are oriented toward certain ways of seeing the rest of the world and thinking about agency and subjectivity. Furthermore, it is not just that these linguistic structures create the world around us; rather, various linguistic structures and vocabularies emerge from different regions of the world in dialogue with the multiple earth others that make up languages.[20] We emerge from biohistorical, technonatural, natural-cultural flows as unique phenomena in a given time and place, and return to affect those flows. Thus we need help rethinking the boundaries of human-nature-machine in ways that might take us beyond the reifications of human exceptionalism—as do both the languages of posthumanism, which uses technology to take us out of the rest of the natural world, and that of romantic environmentalism, which suggests that we have indeed become exceptional and need to return to nature. The language of hyperobjectivity and hypersubjectivity might lead us to transgress the boundaries of human exceptionalism and move us toward a transhuman planetary future.

Hyperobjectivity/Hypersubjectivity

> Could we have a progressive ecology that was big, not small; spacious, not place-ist; global, not local (if not universal); not embodied but displaced, spaced, outer-spaced? Our slogan should be dislocation, dislocation, dislocation.
> TIMOTHY MORTON, *The Ecological Thought*

As Morton suggests in the epigraph to this section, part of the problem with taking materiality seriously is that we have abstracted, somewhat arbitrarily, the boundaries of subjects and objects according to our human, epistemic limits. In other words, what we understand on a commonsense, everyday level to be reality may actually be an abstraction from reality. This is not due to some Cartesian dream from which we will awake, but rather because there is nothing to suggest that boundedness, stasis, and essences are definitive of reality. One thing that is common to most of the scientific, philosophical, and religious theorists that I have discussed throughout this text is that their decision to stick with a metaphysic of presence, or representationalism, was just that: a decision rooted in specific trajectories of biohistorical becoming. Such metaphysics have brought us the technologies that we associate with the industrial revolution, modern Western science, and modernity in general. Again, this modern Western truth regime is no more or less true than any other truth regime. Given the multiplicity of truth regimes that we can live into, all of which are loosed from representational verification, it becomes a difficult task to talk about what progress might mean. Indeed by many measures—such as lower disease rates, lower human-to-human violence per capita than at most points in human history, extended human life span, and even less air pollution than twenty years ago—the modern, Western, mechanical truth regime has led to progress.[21] Such progress may be mostly confined to humankind and at the expense of planetary others, but, nonetheless, by some measurements this is undeniable progress. But we must always question toward what we are progressing: progress always has a goal in mind. Given that there are multiple ways of becoming, and that each relies on certain boundaries of subjectivity and objectivity, defining progress becomes an ethical process rather than a metaphysical assertion. In other words, we can talk of progress just as we can talk of truth, but these categories have become contextual and ethical. Such ethical deliberations and contextual decisions are what we have done and made all along in the face of ambiguity and uncertainty, but now we take responsibility for these faith statements rather than hiding behind false foundations.[22]

From a metaphysic (which is now acknowledged as always already a combined ethics, epistemology, and ontology) of change and becoming phenomena, the boundaries of subjects and objects are understood as permeable. From a planetary perspective that moves beyond humanism and human exceptionalism toward an evolving multiperspectival and multiagential reality; hence, Morton's notion of hyperobjects/hypersubjects becomes quite relevant. For example, we might understand the city as a hyperobject/subject or the process of climate change itself as something that is a hyperobject/subject of which we are a part. Both the aforementioned examples have their own force, their own agency, their own trajectory that is greater than any individual or species ability to control. These phenomena (the city and climate change) are evolving and shaping many planetary phenomena. It is also the case that we might call bioregions and ecosystems hyperobject/hypersubjects, and in this sense James Lovelock's understanding of Gaia as an organism is a forerunner to this concept.[23] However, it would be false to assume that technologies ought not be included and/or that humans are somehow inimical to the body of the earth. This, again, is to assume too much knowledge from our limited human perspective, to stand outside as exceptional, and to diagnose the becoming planetary community as if we were capable of objectivity. Agency in this view is shared, and human agency, much less individual human agency, is but one manifestation of agency among many.

Still another way of understanding objectivity and subjectivity beyond the confines of human exceptionalism is provided by Deleuze and Guattari with their understanding of bodies without organs (or BwOs). To think of subjectivities in particular as something internal rather than as created through a host of external trajectories, organisms, and technologies is to recapitulate the individualism of the metaphysics of presence, stasis, and essence that reinforces human exceptionalism. Through their concept of the BwO, Deleuze and Guattari seek to transgress the boundaries of inner/outer, subject/object, and human/other than human. This is not to be understood as a fragmented self, but rather as the multiplicity of the self.

> It is not at all a question of a fragmented, splintered body, of organs without the body (OwB). The BwO is exactly the opposite. There are not organs in the sense of fragments in relation to a lost unity, nor is there a return to the undifferentiated in relation to a differentiable totality. There is a distribution of intensive principles of organs, with their positive indefinite articles, within a collectivity or multiplicity inside an assemblage and according to the machinic connections operating on a BwO.[24]

Such a BwO includes the virtual flows of histories, biologies, and other organisms that flow into any given subject. In other words, any given manifestation of our subjectivity is always already a conglomeration of the virtual realities that make up that subjectivity. To reify any given moment of subjectivity is to become a cancerous BwO that seeks to reproduce sameness. On the other hand, not to acknowledge the uniqueness of a given BwO is to deny any sort of resistance to the flow of various lines of flight that make up a given BwO. Thus with the notion of a BwO Deleuze and Guattari seek to articulate a multiple understanding of subjectivity that navigates between complete selflessness and complete egoistic individualism. Our internality is nothing without the multiple others with which we are in constant interaction, and our bodies are made up of multiple biological, historical, and cultural others. However one draws the boundaries around a concept or identity, that entity is always already multiple. This also means that entities are always already subject-objects (or subjective objects/objective subjects). It also means that uniqueness is accorded to the various BwOs by virtue of the flows that make up a given embodiment.

As Plumwood notes, there is nothing unethical about recognizing the fact that we all serve as both subjects and objects, nor that we treat others as both subjects and objects. Rather, the ethical problems arise when subjectivity is reduced to the boundaries of an individual (human or species) and others are treated as *mere* objects. Finally, and in close connection to their concepts of becoming plant, mineral, and animal, Deleuze and Guattari's BwOs suggest that any given entity is a mixture of plant, animal, human animal, tissues, technologies, ecologies, imaginings, and agencies. The ontoepistemic cut at the line of an individual subjectivity is mirrored by the ontoepistemic cut at the lines between species, and especially between humans and the rest of the natural world.

Elsewhere, I have suggested that this cut, at least in the historical trajectory of Western thought, is mimicked and supported by the logic of an all-powerful, transcendent God that creates ex nihilo.[25] This ex nihilic logic, which grounds a great deal of Western notions of human exceptionalism and is being globalized around the planet via the regimes of truth known as modern Western science and capitalism, shows up in our legal and social systems. Challenging exceptionalism doesn't just challenge the human/nonhuman divide and our relationships to other animals; rather, it challenges the very foundation of individualism. Since the Lockean liberal individual is at the heart of our economic and legal systems, many changes will be required if we are to shift toward planetary identities. Throughout I have argued for a shift in our understandings of nature, religion, and identity toward an evolving, planetary context. Though I have tried to offer some suggestions for what this will mean for identity politics, environmental ethics, and the concept of human exceptionalism in general, drawing out specific conclusions for what our planetary future might look like is beyond the scope of this book. This is not a cop-out, but a refusal to tie our thought experiments too tightly to ethical, political, and practical outcomes. Moving too quickly toward action denies the reality of the ambiguity of all actions and also plays into the economization of all life on the planet. Outcome-driven thinking can retard the creative process of thinking with the rest of the planetary community through imaginative, possible becomings. To be certain, I am not saying that outcomes don't matter, but I am saying that we need to provide ample space for imagining possibilities. Having said that, I would like to end this chapter and book with some reflections about what moving beyond human exceptionalism might mean for the concepts of agency, hypocrisy, integrity, and love.

Challenging Agency, Hypocrisy, Integrity, and Love

If we understand truth as truth regimes and we understand subjectivities as made up of multiple subject-objects whose boundaries are formed through biohistorical habits of becoming, then we have to understand some of the specific mechanisms by which technologies of meaning operate. To put it (perhaps) more clearly, how is it that we are, for instance, locked into the performance of the Lockean liberal self and its understandings of humans as exceptional to the rest of the natural world? What, on a daily basis—even at emotional and subconscious levels—keeps us bound to the repetition of age-old performances, even when it is clear that such performances are not taking into account many planetary others? Just as we don't usually wake in the morning and say to ourselves, "I am going to be completely different today and forget everything that led up to this point," and then go about abandoning the life we are used to for something completely new, so habits of becoming are not intentionally formed by evil masterminds in an instant. Granted, some evil masterminds do change the world in an instant, and some people do experience complete breaks with their former lives. However, we often understand these as episodes of psychosis or some other lapse in mental health and general well-being. For most of us, change is gradual and qualitative rather than marked by intentional moments of decision. We wake up and ask ourselves, as the David Byrne song does, "Well, how did I get here?"[26]

Philosophers such as Hannah Arendt discuss these gradual changes in terms of the rise of totalitarianism or more generally in terms of the "banality of evil."[27] Little moves begin to add up to something much larger over a period of time. Evolutionary biologists will be familiar with these gradual types of shifts in boundaries as well. It is only in looking back that we might convince ourselves that these changes all make sense together. What we often cover over with our meaning-making practices is the process by which our identities and our understandings of reality are constructed and cobbled together over time. Thus, things *seem* natural, just right, or genuine. There are many constructions or mechanisms that help keep us locked into performing the Lockean understanding of the individual and all that it entails. Here, I will discuss four such constructions/mechanisms: agency, hypocrisy, integrity, and love.

Agency

I have discussed agency throughout the text at greater length than the other three mechanisms that I explore here, but as a recap or reminder, agency is most often conceived in the line of thinking known as Western as something that resides in the individual subject. It is this type of subjectivity that is reflected in the notion of private property (individual labor/agency mixed with dead matter) at the heart of capitalism, which is being globalized. Such agency is reflected in the reified boundaries that lead to the bootstrap mentality of economics so prevalent in the 1980s, which is still prevalent among some fiscally conservative politicians. The idea here is that the buck, quite literally, stops with the individual subject: regardless of the biohistories that shape class, race, gender, sex, sexuality, and embodiment in general. In other words, such an understanding of agency backgrounds the reality of unearned privilege (based on class, race, gender, sex, sexuality, and ableism) and the "slow violence" of environmental racism and injustice.[28] Such individualistic agency is not only reflected in the economics of "free" market capitalism but also in the Western legal system based upon individual rights.

Humanism and its corresponding (Lockean) liberal understanding of humanity has done much to further the endeavors of individuals. However, such an understanding of subjectivity and individualized agency serves to background many forms of social and ecological violence. We might look at the rate of imprisonment of black males in the United States versus the imprisonment of their white counterparts, for example. Speaking personally, I know that as a wild teenager growing up in Little Rock, Arkansas in the 1990s my transgressions were often overlooked because of the whiteness of my skin. In other words, had I been born black or even a darker-skinned Latino, I would have likely gone to prison for some of the rowdy teenage stunts I pulled rather than to the best liberal arts college in the state of Arkansas. My subjectivity is shaped by biohistories that are far beyond my control or influence. My subjectivity and my agency are then not necessarily my own. How is it, then, that responsibility is always narrowed to the individual subject when it comes to legal and economic institutions? Such a notion of responsibility is hubris if not a narcissistic fantasy: as if people or any one person *could* have that type of power.

Distributive justice is rarely considered as a viable basis for legislation in this country. How do we begin to set up a legal system that hinges on the concept of shared agency, fuzzy subjects, and identities (in general) that are always already informed by others and transgressing boundaries established by present manifestations of becoming? Again, international laws such as those that call on one-fifth world nations to pay more for the consequences of climate change than four-fifths world nations, precisely because it is the one-fifth that has benefited from the unbridled use of fossil fuels, are laws that attempt to recognize some notion of shared agency. These laws are often highly contested by those that still perform and believe in liberal humanism. Finally, there are rarely laws that transgress the sacred boundary of human exceptionalism: animals are always understood as moral patients at best (never as moral agents) and resources for human use at worst. Little account is given to the sustenance that makes up the embodied existence of so-called individual humans, much less to the biohistories of evolution that have led to the emergence of *Homo sapiens*. For the most part, the only time a nonhuman animal is treated as an individual by law is in the case where an animal kills or harms a human being: for this that animal must pay with its life. Our legal system is mostly set up to recreate the boundaries of individual human beings over and against any sort of notion of collective, planetary becoming.

Hypocrisy

A second mechanism that helps keep us locked into exceptional thinking and the corresponding notion of individualism is the logic of hypocrisy. Hypocrisy works as a social control to help keep us locked into consistent performances of identity. It is akin to representationalism in epistemology, which seeks correspondence between our understanding and reality as it truly is. In this case, however, past manifestations of our identities are held in stasis as our true or genuine identity and all of our actions are judged against this static identity. We are quite literally shamed by social pressures into performing a consistent self. What happens if we exist as becoming subject-objects who are always transgressing the boundaries of our identities? Furthermore, what happens when we live in an entire planetary community that is always in the process of becoming? How are we to change with, rather than try and block, the changes of a becoming reality? The threat of hypocrisy might shame someone into submitting to business as usual simply because it is easier than dealing with the messy process of transgressing boundaries toward different future realities. I am by no means saying that there is no legitimate, negative connotations of hypocrisy; rather, I am merely suggesting that the very fear of being a hypocrite often bullies us into reified versions of ourselves.

Consider that we are all always already a mixture of good and bad or that each and every action always has positive and negative consequences. In the context of global climate change it is impossible for any of us to live without adding to a carbon footprint. There are things we can do to reduce that footprint, but we can't live as if we do not contribute to the planet's carbon load. This means, to a certain extent, embracing the hypocrisy of our individual becomings. In other words, we have to live with the fact that our subjectivities contain many conflicting realities and that we also change over time. Furthermore, it means realizing that in order to live into an ideal identity one must have the economic and legal power to create such an identity. To move back to the land, always buy organic and local, never fly (which few self-proclaimed environmentalists can avoid), choose to always eat vegetarian or vegan, and/or to choose to pay more for clothes and other products that are produced on a smaller scale without the use of unknown sources of global labor are all benefits of certain classes of people. To be sure, many people in the four-fifths world outside of the West live off the economic grid in a much more sustainable way than those who can afford the haute couture of Western environmentalism. Yet these people are often seen as impoverished ethical patients rather than as outstanding, exemplary global citizens. In any event, hypocrisy can be used to keep people on the capitalist, liberal humanist treadmill by feeding one's desires to make enough money so that one can afford solar panels and all-organic and all-local products. The uncomfortable tension created by caring about these things and yet not being able to live completely according to one's ideals combines nicely with the fear of being a hypocrite and serves as a powerful mechanism to keep us locked into our individualistic ways of thinking. Since we are all incapable of living up to impossible ideals, the fear of hypocrisy helps keep human becomings in the realm of the status quo, business as usual, and normalcy.

In order to break out of these mechanisms of human exceptionalism, liberal individualism, and global impositions, perhaps planetary identities would be better served by recognizing our inherent hypocrisy. We must cobble together strange ethics, odd ideals, and inconsistent truths in order to create new habits for planetary becoming. We must transgress our human identities, our subjectivities, and our notions of what is consistent if we are ever to reveal the cracks in the façade of a global reality. Such cracks also take place in our understandings of integrity.

Integrity

What does it mean to be an integral person? In a sense, it means avoiding hypocrisy, and thus some of the same challenges can be presented to the concept of integrity. In a deeper sense, though, living an integral life means living a complete life (vocationally, socially, environmentally, interpersonally, and personally) that is congruent with the principles in which you believe. Again, a certain socioeconomic status is required to background all the inconsistencies inherent in living life. Destruction and death are surely a part of every process of cocreation and becoming. Integrity from a global perspective is only possible if one imposes one's ideals on the face of the globe and backgrounds the "dark" underside of the consequences of such an imposition, similar to the process, discussed in earlier chapters, of how the Enlightenment cocreated the Dark Ages. In a certain sense, the concept of integrity serves to reinforce foundational, representational, substantial, and generally static thinking. From an evolving planetary context of multiple becomings, integrity might be redefined as contextual ethical thinking with planetary others on the move. When one encounters and realizes her continual coconstructedness along with multiple planetary others, integrity becomes a mobile concept that is concerned with honoring the multiplicity of subjects. It is, in other words, OK to be a completely different person from the person you were twenty years ago. Furthermore, it is the case that multiple scalar analyses reveal that integrity is not only a located concept in terms of evolving subjectivities in space-time, but it is also a scalar project that could benefit from metaphors/mechanisms like hyperobjectivity/hypersubjectivity and BwOs. From these perspectives, we might ask whether we are talking about the genetic, chemical, animal, plant, molecular, physical, psychological, or planetary level of life when we pose the question of integrity. One's own integration is at the expense of the disintegration of multiple earth others, which some Jain monks who intentionally fast themselves to death no doubt realize. Negotiations between these multiple levels of planetary integrity will no doubt have to be waged in the arena of politics rather than under the false premises of assumed human exceptionalism, which is always already about economics and politics. We will no doubt have to tease out what we mean by planetary loves: that which we want to see thrive in the future becoming of the planetary community.

Multiple Planetary Loves

David Abram writes: "Gravity—the mutual attraction between our body and the earth—is the deep source of that more conscious delirium that draws us toward the presence of another person. Like the felt magnetism between two lovers, or between a mother and her child, the powerful attraction between the body and the earth offers sustenance and physical replenishment when it is consummated in contact."[29] Though polyamory of place and multiple planetary loves were discussed in chapter 6 as ways of transgressing the capitalist politics of love, here at the end of this book, love will get the final word. How do we begin to transgress human exceptionalism and our own individualistic subjectivities in order to transgress our anemic humanistic condition without developing love for that which we cannot know? We have been trained to narrow our emotions through various mechanisms so that they begin and end at the boundary of our present embodiment. The evolutionary biohistories and multiple lives that lead to the emotions we now experience are forgotten for the certainty and satisfaction of a definite and immediate love of a stable, individual, unchanging self. Multiple religious writings, philosophies and artists, writers, and others speak of the necessity of nonattachment as a prerequisite for love. Such nonattachment means different things for different types of love, and I want to end here with a discussion of a few types of love, all of which are founded on nonattachment.

First, there is the erotic love that comes with the phenomenon of embodiment. Such love from a global context adheres to the capitalist politics of love mentioned in chapter 4. It is selfish and always *for* the stabilization of a static self rather than for the flourishing of the other(s). This type of erotic love from a planetary perspective means love for the health, well-being, growth, and continual evolution of an embodied other. Such an other can never be possessed or fully known, for he/she/it is always in the process of becoming. It is in this sense that he/she/it can only be loved in recognition of his/her/its otherness. This is not an otherness that ignores the always already of intertwined subjectivities in embodied becoming, but is an otherness of the unknown possibilities for earth others' becomings. Transgressive love in the planetary erotic sense does not possess any given manifestation of self or other. Furthermore, its ramifications go beyond the current subjectivities involved and are *for* the planetary community.

A second type of love is the more filial love of and for the planetary community. This is the love that will begin to allow us to transgress our epistemontologies of human exceptionalism. Such a love, nonpossessive and nonattached, is *for* the flourishing of multiple earth bodies. This is the type of nonviolent love that is found in many of the world's religions, but especially in the sense of harm reduction. As such, we ought to use information we gain from animal studies, ecologies, medicine, biology, chemistry, climatology, psychologies, and the multiple other ontoepistemic levels of the planetary community in order to work toward this reduction of violence on the planetary scale. We do not have to be beholden to an orthodox interpretation of the laws of karma, for instance; but rather can transgress and per/vert them given the multiple perspectives of the planetary community. In other words, karma might be rethought in terms of taking multiperspectivalism seriously rather than reducing all laws of karma to the goal of liberation, which can only (traditionally) be attained from an embodied location as human. Such a filial love for the planetary community will provide a basis for transgressing the present truth regime and moving in alternative lines of flight toward the planetary future.

It is only here at the end of this embodied textual journey that we might be able to begin to speak of a viable, agnostic, and planetary *agape* or divine love. Lest the reader think I imagine these three loves as mutually exclusive, I do see them as intertwined. In fact, they all depend upon one another and mutually influence one another. I could obviously start with an examination of the construction of these three different understandings of love, beginning with the Greeks, but let me just suggest that the genealogy and development of these three loves at this point in biohistory intersects the biohistories of Greek trajectories, many religious traditions (including the Abrahamic faiths and Vedic traditions), philosophical thought (including poststructural and queer understandings of otherness and transgression of boundaries), and much contemporary science (including information from physics, evolutionary biology, and the neurosciences). In other words, it is a love that is cobbled together by the fruits of this book. Just as the erotic-planetary love is not simply *for* star-crossed lovers nor simply an economic or social arrangement but is rather *for* the planetary community; and, just as the filial understanding of planetary love transgresses species boundaries and is *for* the future becoming planetary community, so this understanding of *agapic* planetary love is *for* the planetary community.

Many religions, not to mention iconoclastic and deconstructive philosophies, have jolted us out of the habits of love based upon ego, family, kin, and even nation. The cry of Jesus to leave one's family and create a new family not based on kinship, race, or even sex; the Buddha's rejection of the more self-concerned ends of both wealth and asceticism; and the yearning for human, animal, and earth liberation from oppressions all provide fertile grounds for imagining alternative planetary futures. These imaginings are not about transcending, but rather are immanent phenomena of thinking with the planetary community toward new habits and apparatuses for becoming. *Agapic* love is the nonattached type of love to planetary imaginings. Much like Kaufman's understanding of the imagination in the role of theology, this understanding of love is that which grounds the possibility for a nonattached love in the first place.[30] Though a recent emergent phenomena within the planetary community, such love may be the grounds for planetary becomings that transgress global impositions of sameness.

This type of love hinges on a type of viable agnosticism and ambiguity toward our continuous planetary becoming. In a world where subjectivities are always already multiple, hybrid, and transgressing boundaries, there is no such thing as a complete other. Rather, the space held open for the other is the space for the unknown, the space in which creative possibilities emerge (not ex nihilo but ex profundis).[31]

It is this space, not blank but full of possibilities and ever transformed in the process of becoming, that might be called divine creativity or mystery. Such a notion of emptiness, or that which is not ordered or visible, is important to Deacon's understanding of the possibility for emergence as well. For

Deacon it is the space at the center of the spokes of a wagon wheel that makes all the difference in the functioning of the wheel; similarly, it is the unknown, that which is left open or left out, that allows for the possibility of creative planetary becoming and truly loving the multiplicity of planetary others. This is the open space at the center of the wagon wheel necessary for its function, this is the space in between all hybrid *post-* thought, this is the space of the abject necessary for subjective identities, and this is the negative move in apophatic and deconstructionist modes of thought that release possibilities for growth and change. It is also this ambiguity and agnostic stance toward our own subjectivities that allows for self-love rather than an imposition of closure from some idealized understanding of ourselves.

Such an understanding of mystery, of God, of divinity is a reclaiming of the "god of the gaps." Those in the science and religion dialogue often understand this phrase as denigration of religion in some way. In other words, the idea is that science will one day explain away the need for religion. However, I have argued here that such a reductive model denies the role of humans as meaning-making creatures and also adheres to a metaphysic of substance and an epistemology of representationalism. Built into both science and religion is a sense of the unknown and ever changing process of becoming life: the truth regimes of science and religion are always changing and changing the worlds in which we live. When these truth regimes are taken to be reality *en toto* or as closer to nature *en esse*, then violence is perpetuated on the becoming planetary community. This violence is the violence of possession masked as love, and the love of reifications is destructive to the planetary community. To love, then, is to embrace the evolving and necessary unknowability of all of our thoughts, ethical justifications, imaginations, hopes, dreams, values, and knowledge. In fact, the unknowing of the world has always played an important tricksterlike role in the becoming of new ways of being in the world, for better and worse. The only certainty is that when certainty is imposed on the world love is impossible and violence is inevitable. Hence at the very center of the emergence of planetary identities is an embrace of the god of the gaps, the cloud of unknowing that is the source for the continuation of the becoming planetary community.

NOTES

INTRODUCTION

1. Postcolonial studies abound with metaphors of hybridity, boundary thinking, interstitial identity formation, etc. All these tropes aim to identify the experience of this hybrid reality in which we live and thus challenge the notion of the isolated individual self popular with Western monotheisms and philosophical reflection. See, e.g., Bhabha, *The Location of Culture*.
2. Haraway, *When Species Meet*.
3. Foucault, *Power/Knowledge*, 131.
4. Brennan, *Globalization and Its Terrors*.
5. Barad, *Meeting the Universe Halfway*, 137.
6. I use the "one-fifth" world here to mark economic difference in the world's human population. In other words, if you split the world into five groups of 20 percent, those of us born in the U.S. and Europe are in the top fifth in terms of economic wealth and resource consumption. This means that roughly 80 percent of the world's population consumes at lower levels, yet 80 percent of the world's population must also deal with the effects of consumption due to the top 20 percent. The first data on the "champagne economy" can be found in the 1992 report of the United Nation's Human Development Project: http://hdr.undp.org/en/reports/global/hdr1992/. Though the report is from 1992, the 2010 report suggests the same types of gaps between the one-fifth and four-fifths worlds, despite overall increases in the Human Development Index.

7. Karen Barad (quoted earlier) is quite fond of referring to the *Gedanken* experiments used by earlier physicists such as Bohr and Einstein to work through problems without necessary practical results. However, thought experiments that don't seem to have practical results at a given time end up producing many effects that can later be measured.
8. On the difference between "grounds" versus "foundations" see Keller and Kearns, "Introduction: Grounding Theory" in their *EcoSpirit*, 1–20.
9. Heidegger, *The Question Concerning Technology*.
10. Spivak, *Death of a Discipline*, 73. I draw out my understanding of "planetarity" based upon Spivak's metaphor laid out in this text.
11. Val Plumwood describes the violence resulting from conceptual "backgrounding" in *Environmental Culture*.
12. Mignolo, *The Darker Side of the Renaissance*.
13. I will draw from Karen Barad's metaphor of a phenomenal-based understanding of reality throughout the text. However, many other thinkers offer such an understanding of the world, and these thinkers have deeply influenced my own understanding: Emergent theorists, Process thinkers Deleuze and Guattari, Donna Haraway, and Bruno Latour, among others.
14. Spinoza, *Ethics*.
15. Carolyn Merchant's trope "death of nature" refers to the very process by which modern Western science codes nature as "dead matter" and thus as in a category for use by human beings. Merchant, *The Death of Nature*. The "death of nature" I mean to invoke here is the death of the concept of nature as apolitical, pure, pristine, and somehow separate from culture. See, e.g.: Morton, *Ecology Without Nature*; and Latour, *Politics of Nature*.
16. I take the metaphor of "evolutions rainbow" from: Roughgarden, *Evolution's Rainbow*.
17. This, by the way, is the greatest (mis)reading of Feuerbach's notion of religion as projection. He never then reasoned that religions didn't matter, but rather that they mattered the world around us and should be taken seriously. To locate meaning-making practices within the realm of human histories and cultures is not to declare religion as passé, but rather to suggest that these emergent features of our biohistorical evolution are not imposed from "out of this world." They are *of and for* this world and they matter a great deal.
18. For a brief discussion of emergence theory and its relevance for meaning-making practices, read Goodenough and Deacon, "The Sacred Emergence of Nature."
19. On "lines of flight" see Deleuze and Guattari, *A Thousand Plateaus*.
20. See, e.g., Bauman, "Religion, Science and Nature."

21. Such planetary forms of love are articulated in many of the chapters of Moore and Rivera, *Planetary Loves*.
22. The term *polydoxy* is offered as a direct refutation to orthodoxy. In other words, multiple interpretations are not bad and in fact increase possibilities for future becoming. Furthermore, there has never been a single (orthodox) interpretation of any history, religion, culture. On this concept see Keller and Schneider, *Polydoxy*. Nomadology is again a metaphor of Deleuze and Guattari. It suggests identities of movement and change rather than substance, essence, and stasis. See Deleuze and Guattari, *A Thousand Plateaus*.
23. See ibid.
24. Hefner, *Technology and Human Becoming*, 5.
25. Morton, *The Ecological Thought*.

1. Religion and Science in Dialogue

1. See, e.g. Merchant, *The Death of Nature*; and Dupre, *Passage to Modernity*. Taylor, *A Secular Age*.
2. On these various models for relating "science" and "religion," see Barbour, *Religion and Science*.
3. Horkheimer and Adorno, *Dialectic of the Enlightenment*.
4. For a full examination of the way in which foundational assumptions shape colonizing attitudes toward human and earth others, see Bauman, *Theology, Creation and Environmental Ethics*.
5. For a succinct definition of logocentrism, see Derrida, *Of Grammatology*, 3.
6. Nancy also argues along these lines in *The Creation of the World or Globalization*.
7. For an excellent account of the "wonder" or unknowing at the base of all knowledge, see Rubenstein, *Strange Wonder*.
8. On grounds versus foundations, see Catherine Keller, "Talking Dirty: Ground Is Not Foundation" in Keller and Kearns, *EcoSpirit*, 63–76.
9. Rubenstein, *Strange Wonder*, 45.
10. Ruether, *Gaia and God*, 254.
11. Haraway develops the concept of nature-culture. See, e.g., *Simians, Cyborgs, and Women*, 149–182. I take the concept of biohistory from Kaufman, *In Face of Mystery*.
12. Heidegger, *The Question Concerning Technology*. I only say "modified" here because Heidegger seems to indicate that we might break out of enframing in order to get to some experience of the "ground of being." I would argue that any such "breaking out of" would be an attempt to thwart culture/language/idea for some sort of genuine experience of nature/material/body (or vice versa).
13. Durkheim, *The Elementary Forms of Religious Life*.

14. Tweed, *Crossing and Dwelling*.
15. In terms of environmentalism as a religion, see e.g., Taylor, *Dark Green Religion*.
16. Such integral methods are suggested by Hargens and Zimmerman in *Integral Ecology*.
17. For an understanding of a multiperspectival reality, see Harding's concept of "strong objectivity" in Harding, *Is Science Multicultural?* Haraway's understanding of "situated knowledge" in Haraway, "Situated Knowledges"; and Deleuze and Guattari, *A Thousand Plateaus*.
18. Barad, *Meeting the Universe Halfway*.
19. Griffin, *The Reenchantment of Science*.
20. More on this will be said in later chapters, but I also discuss planetary technologies in Bauman, "Technology and the Polytheistic Mind."
21. Kevin O'Brien develops a type of multiscalar ethics in O'Brien, *An Ethics of Biodiversity*, 108.
22. See, e.g., Goodenough and Deacon, "The Sacred Emergence of Nature."
23. Spinoza, *Ethics*, 20–22. Again, *natura naturans* refers to "nature naturing" while *naturata* refers to "nature natured." The former is verb and the latter is noun.
24. The use of "creative-destructive" here should not be confused with the use of "creative-destruction" by some economists and other scholars to describe the process of capitalism.
25. Bhabha, *The Location of Culture*.
26. Goodchild, *Theology of Money*, 32.
27. On a side note, astrology is itself a perfect example of the efficacy of meaning-making practices. In order to believe that astrology is really REAL, one must still believe in this realm of fixed stars and the geocentric view of the universe. However, most people who get readings today, I argue, do not believe in this sort of world, yet find meaning in such readings. Further, these meanings matter: in the way he lives out a day, interacts with others, etc. Thus, astrology is real in that it effects material-energy relations within and between living organisms on the planet.
28. Note that this is also a bit more abstract than the "three-tiered" cosmologies found in the Ancient Near East, in Indic traditions such as Jainism, or in many indigenous traditions. The three-tiered cosmology was based on the embodied sensory experience of being "in the world" (there is an above, a middle where we stand, and a below). The Aristotelian cosmology begins to rely on technologies of mathematics and optics to extend our senses to imagine the worlds in which we live. It is a mistake to assume that such an extension, whether in Ancient Greece, the Golden Age of Islam, or contemporary times, is a transcendent description

of the "way things are." Rather, we might think of these various technologies of understanding as regimes of truth that cocreate the worlds around us into certain ways of becoming.
29. Mignolo, *The Darker Side of the Renaissance*.
30. Ibid.
31. The colonizing tendencies of Western science, along with its gendered implications, are described well in Scheibinger, *Nature's Body*.
32. For these comparisons, see Noble, *The Religion of Technology*.
33. For a discussion of *globalatinization* in the way I understand it here, see Derrida and Vattimo, *Religion*, 178.
34. Calhoun, Juergensmeyer, and van Antwerpen, *Rethinking Secularism*, 8.
35. Taylor, *A Secular Age*, 3.
36. Charles Taylor as quoted by Jose Casanova in Calhoun, Juergensmeyer, and Vanantwerpen, *Rethinking Secularism*, 59.
37. Connolly, *A World of Becoming*, 38.
38. Latour, *Politics of Nature*, 187.
39. Asad, *Formations of the Secular*, 13.
40. Ibid., 73.
41. Althaus-Reid speaks to the liberating quality of multiple (per)versions in her *Indecent Theology*.
42. *Différance* is that which is always leftover in any concept or construction of the world. It is a term developed by Derrida to suggest that there is always a remainder in any linguistic, conceptual, or other description of reality or other. There is no "god's eye" view of an other (or even the self) and thus everything we use to describe self, other, world, things, etc. will always include a left out "remainder," or *différance*.
43. Weber, *The Protestant Ethic and the Spirit of Capitalism*; Dupre, *Passage to Modernity*. For a fuller description of this, see Bauman, *Theology, Creation, and Environmental Ethics*.
44. Merchant's book *Reinventing Eden* describes this process well, see pp. 75–79.
45. See ibid.
46. Asad, *Formations of the Secular*, 192.

2. Destabilizing Nature

1. *Misplaced concreteness* is a term used by the process thinker Alfred North Whitehead to describe the way in which we tend to attribute stability and certainty where there is none. Our language often mistakes the constant flux of nature (or what Spinoza calls nature naturing) with a stable reality. Assuming stability in anything living is always an instance of "misplaced concreteness."

2. Žižek, In Defense of Lost Causes, 446.
3. Though Teilhard offers an amazing, evolutionary understanding of salvation that breaks down many dualisms—particularly between humans and nature—it is still a salvation narrative that takes place between ultimate origin and end, in this case Jesus Christ, the Omega Point of the Universe. Such a narrative collapses all otherness into a single (Hegelian) History. See de Chardin, *The Phenomenon of Man*.
4. The universe story as articulated by Thomas Berry and Brian Swimme offers another amazing revolution in thinking of humans as part of the expanding universe and as part of the ongoing process of evolution, yet they understand humans as the universe thinking about itself. Such a move seems to me a bit too anthropic. Rather, what we need are perhaps multiple stories of the universe or multiverses. See e.g., Swimme and Berry, *The Universe Story*; and the more recent book (and film) by Swimme and Tucker, *Journey of the Universe*.
5. Foucault, *The History of Sexuality*, 86.
6. Again I am taking up the notion of radical immanence primarily from Deleuze and Guattari, *A Thousand Plateaus*.
7. Morton, *The Ecological Thought*, 40.
8. Weststeijn, "Spinoza sinicus."
9. Barad, *Meeting the Universe Halfway*, 137.
10. Ibid., 140.
11. On the notion of "strong objectivity" that understands the subject is a part of the object being observed, see Sandra Harding, "Rethinking Standpoint Epistemology: What Is Strong Objectivity," in Alcoff and Potters, *Feminist Epistemologies*, 49–82.
12. See, Merchant, *The Death of Nature*.
13. See my article, Bauman, "Fashioning a Persuasive Environmental Ethic."
14. See, McKibben, *The End of Nature*.
15. I follow Timothy Morton in this line of argumentation; see Morton, *The Ecological Thought*, 125ff.
16. On "postmodern sciences" see Griffin, *The Reenchantment of Science*. Postmodern does not indicate here a loss of Reality. Most scientists still assume they are studying Reality. Rather, it means an acknowledgment that this reality is more about energy events rather than substances and that even the way we approach reality, to some extent, codetermines what we will "see."
17. Singer, "The Cosmology of Giordano Bruno," 192. Cf. her biography, *Giordano Bruno*.
18. See, e.g., Clarke, *Oriental Enlightenment*.

19. Campbell, *Wonder and Science*, 120.
20. Bruno is (again) indebted to Cusa, and much has been written on Cusa and deconstructionism, polydoxy, and the importance of "unknowing" at the heart of knowledge. I am particularly indebted to Catherine Keller's work on this figure. See, e.g., Catherine Keller, "The Cloud of the Impossible: Embodiment and Apophasis" in Boesel and Keller, *Apophatic Bodies*, 25–44.
21. Note: Bruno was, of course, famously, burned at the stake, but not necessarily for his "scientific" ideas. John Hedley Brooke, for instance, argues that it was much more because of his heretical ideas about Jesus than anything else. In any event: there was no "religion" versus "science" to speak of in Bruno's time, so the projection of that onto his trial is at best a faulty anachronism that perpetuates the idea that they are mutually exclusive ways of knowing. See, e.g., Brooke, *Science and Religion*, 39–40. Cf. Numbers, *Galileo Goes to Jail*, 59–67.
22. See, e.g., Spinoza, *Parallel Passages*.
23. Spinoza, *Ethics*, vii.
24. "For Spinoza, there is one immanent substance, and human being is a mode of the attributes of nature—thought and extension," in Gatens, "Feminism as 'Password,'" 60.
25. Spinoza, *Ethics*, part 1, P:29.
26. Weststeijn, "Spinoza Sinicus."
27. Eckstein, "The Religious Elements in Spinoza's Thought," 162.
28. Here I agree with the pragmatists: "For pragmatists, there is no such thing as a nonrelational feature of X, any more than there is such a thing as the intrinsic nature, the essence, of X." Rorty, *Philosophy and Social Hope*, 50.
29. Deleuze and Guattari, *A Thousand Plateaus*, 3–25.
30. Spinoza, *Ethics*, part 1, appendix 1. Spinoza does not take any sort of transcendent foundationalism well. He writes, "And so they will not stop asking for the causes of causes until you take refuge in the will of God, that is, the sanctuary of ignorance" (ibid.).
31. Stephen R. L. Clark, "Pantheism" in *Spirit of the Environment*, 45.
32. Dorothea Olkowski, "Political Science and the Culture of Extinction" in Herzogenrath, *Deleuze/Guattari and Ecology*, 153.
33. Deleuze, *Expressionism in Philosophy*, 255.
34. Bergson, *Creative Evolution*, 89.
35. Ibid., 248.
36. Bergson, *The Two Sources of Morality and Religion*.
37. There is a lot of crossover in these three different philosophical endeavors, which are, at least, a) radical immanence, b) humans as meaning-making creatures, and

c) an open future. There is also a lot of crossover in the annotations of works in these camps. For instance, Terrence Deacon relies heavily on Peirce (not to mention Confucianism); Deleuze and Guattari rely on the *élan vital* of Henri Bergson, a predecessor to the "emergence" discussion today. All of them draw from Spinoza. Yet, they do not seem to be having present-day conversations. Finally, they all belong to a postfoundational type of tradition that might also be called antijuridical. See, e.g., Gatens, "Feminism as 'Password,'" 60.

38. Hargens and Zimmerman, *Integral Ecology*, 141. Cf. the development of the concept of habitus in a Bourdieu, *The Logic of Practice*, 52–65.
39. Rorty, *Philosophy and Social Hope*, 38.
40. Bruno Latour, "Thou Shalt Not Freeze-Frame" in Proctor, *Science, Religion and the Human Experience*, 46.
41. Barad, *Meeting the Universe Halfway*, 234.
42. Rue, *Nature Is Enough*.
43. Deacon, *The Symbolic Species*.
44. Deleuze and Guattari, *A Thousand Plateaus*, 189–190.
45. Nancy, *Being Singular Plural*, 3.
46. Deacon, *Incomplete Nature*, 3.
47. Ibid., 264–287.
48. Deleuze and Guattari, *A Thousand Plateaus*, 232–309.
49. Deacon, *Incomplete Nature*, 539–540.
50. Said, *Orientalism*. Its not necessary as that is the general argument of the whole text and one of the major teneants of the concept of orientalism.
51. Kaiser, *How the Hippies Saved Physics*, 3.
52. Schneider and Sagan, *Into the Cool*, 82.
53. Barad, *Meeting the Universe Halfway*, 203.
54. Ibid., 315.
55. Ibid., 361.
56. See, e.g., Gudorf, "The Erosion of Sexual Dimorphism."
57. Barad, *Meeting the Universe Halfway*, 379.
58. Butler, *Bodies That Matter*.
59. Roughgarden, *Evolution's Rainbow*.
60. Althaus-Reid, *Indecent Theology*.
61. Barad, *Meeting the Universe Halfway*, 335.
62. Derrida, *Writing and Difference*, 161.
63. See Latour, *Politics of Nature*.
64. Bhabha, *The Location of Culture*.

65. Schneider and Sagan, *Into the Cool*.
66. Foucault, *Power/Knowledge*, 109–133.
67. On the repoliticization of "nature," see Latour, *Politics of Nature*; and Morton, *Ecology Without Nature*.

3. Destabilizing Religion

1. Bruno Latour, "Thou Shalt Not Freeze-Frame" in Proctor, *Science, Religion, and the Human Experience*, 36.
2. Again, the metaphor of "lines of flight" is taken from the work of Gilles Deleuze and Felix Guatari.
3. For a great analysis on the language of "dark energy" and "dark matter" as it relates to "light supremacy" and racism, see Holmes, *Race and the Cosmos*. This is, of course, a topic of much postcolonial analysis as well, e.g., Mignolo, *The Darker Side of the Renaissance*; and Keller, Nausner, and Rivera, *Postcolonial Theologies*.
4. Lewellyn, *Margins of Religion*.
5. Keller, *God and Power*, 149ff.
6. Goodchild, *Theology of Money*, 32.
7. Ibid., 53.
8. Lorde, *Sister Outsider*, 110–113.
9. Bauman, *Globalization*.
10. See, e.g., Boesel and Keller, *Apophatic Bodies*; and Keller and Schneider, *Polydoxy*.
11. See, e.g., Bauman, *Theology, Creation, and Environmental Ethics*.
12. Hyde, *Trickster Makes This World*, 7.
13. For a collection of some Native American trickster stories, see Jackson, *The Wisdom of Generosity*.
14. Anzaldúa, *Borderlands*.
15. De la Torre and Hernandez, *The Quest for the Historical Satan*, 206ff.
16. For a good historical and philosophical analysis, see, e.g., Bedau and Humphreys, *Emergence*.
17. Though this is not uncontroversial. Some would argue that there is no need for "top-down" causation at all. I argue, however, that a philosophically robust account must provide some sort of suggestion for how and why ideas matter to the world, or else they are reduced to "lower levels" of reality.
18. Clayton, "Conceptual Foundations of Emergence Theory," 2.
19. For a good discussion of this, see Haag and Bauman, "De/Constructing Transcendence."
20. Clayton, "Conceptual Foundations of Emergence Theory," 2.

3. DESTABILIZING RELIGION

21. For a discussion of "ethical" and "epistemological" anthropocentrism, see Plumwood, *Environmental Culture*, 167–195.
22. Bellah, *Religion in Human Evolution*, 34.
23. Gilles Deleuze and Felix Guattari seek to evoke such a metaphor in the "rhizome" and Bruno Latour in "the collective" (see discussion of both further on in this volume), but I think it is just as important to evoke these embodied ways of becoming in the world using more sensuous metaphors. In other words, what does it *feel* like to embody the "rhizomatic" thought of Deleuze and Guattari or the collective process that Latour describes?
24. Harding, *The Postcolonial Science and Technology Studies Reader*, 34.
25. Merchant discusses the problems with such narratives in *Reinventing Eden*.
26. Kaufman, *In the Beginning*, 44.
27. "Our *historicity* . . . is the most distinctive mark of our humanness." Ibid., 45. See also Kaufman, *In Face of Mystery*, 117.
28. Ronald Bogue, "A Thousand Ecologies" in Herzogenrath, *Deleuze/Guattari and Ecology*, 49–50.
29. I follow a metaphysics of immanence from Bruno to Spinoza and to the American pragmatists, the theory of Emergence, and French "post" thinking such as is found in Gilles Deleuze and Felix Guattari's work.
30. "Nearly all [Christian groups] accepted the basic schema which elaborated a conception of God, and of God's Truth, as having independence and objectivity over against humanity." Kaufman, *An Essay on Theological Method*, 28. See also: "Christians may no longer consider themselves to be authorized in what they say and do by God's special revelation." Kaufman, *In the Beginning*, 68. In fact, Kaufman also argues correctly that idealism and materialism are the same thing: "Materialisms themselves are, thus, at once products and examples of *spirit*, in the (empirical) sense in which I am using that word here." Kaufman, *In Face of Mystery*, 259.
31. Kaufman, *An Essay on Thelogical Method*, 47.
32. See, eg., Foucault, *Power/Knowledge*.
33. Kaufman, *In Face of Mystery*, 67.
34. Keller, *Face of the Deep*. Cf. Bauman, *Theology, Creation, and Environmental Ethics*. Many of the ideas in this chapter began to take form in that book.
35. Kaufman, *An Essay on Thelogical Method*, 40.
36. Ibid., 8.
37. Another, similar, way to think of it is as what Lorainne Code describes (following Castoriadis) as "social imaginaries": "Imaginatively initiated counterpossibilities [that] interrogate the social structure to destabilize its pretensions to naturalness and wholeness, to initiate a new making (a *poesis*)." Code, *Ecological Thinking*, 31.

38. Kaufman, *An Essay on Thelogical Method*, 34.
39. Van A. Harvey's book is key for those who want to reread Feuerbach beyond the straw interpretation of "God as mere projection." Van Harvey, *Feuerbach and the Interpretation of Religion*.
40. Sabatino, "Projection as Symbol," 183.
41. Green, *Theology, Hermeneutics, and Imagination*, 92. Furthermore, Green writes, "So the scholar of religion must say to Feuerbach: Yes, the imagination is indeed the source of religion, but No, religion is not thereby disqualified from the search for truth" (ibid., 103).
42. Keller, *Face of the Deep*, 219–220.
43. Keller, *God and Power*, 118.
44. Plumwood discusses what a participatory democracy might look like in *Environmental Culture*, 93–96.
45. Rorty, *Philosophy and Social Hope*, 152.
46. See, e.g., Deleuze and Guattari, *A Thousand Plateaus*, 25: "Making a clean slate, starting or beginning again from ground zero, seeking a beginning or a foundation—all imply a false conception of voyage and movement . . . [rather we proceed] from the middle, through the middle, coming and going rather than starting and finishing."
47. To offer ground is not to make a foundational claim but to "give reasons, to cede turf, and to remember the shared earth that provides the one common ground in which all of our contexts nest." Keller and Daniel, *Process and Difference*, 13. See also Keller's chapter, "Talking Dirty: Ground Is Not Foundation," in Keller and Kearns, *EcoSpirit*, 63–76.
48. See, e.g., Catherine Keller on John Cobb's concept of "the common good": "The university has organized knowledge/power according to standards of universality and objectivity that mask the special interests of race, class/economics, sex/gender, and species." Catherine Keller, "Process and Chaosmos: The Whiteheadean Fold in the Discourse of Difference," in Keller and Daniels, *Process and Difference*, 55.
49. It is beyond these horizons that Kaufman's concept of "mystery" is helpful: "It is in terms of that which is beyond our understanding that we must, finally, understand our human language." Kaufman, *In Face of Mystery*, 6.
50. "Rationality is about responsibility: the responsibility to pursue clarity, intelligibility, and optimal understanding as ways to cope with ourselves and our world." Van Huyssteen, *The Shaping of Rationality*, 2.
51. From within this theology as conversation model, reason, according to Kaufman, is "not . . . a kind of reservoir or bank from which our moral rules and principles

184   3. DESTABILIZING RELIGION

can be withdrawn as needed; it is, rather, simply our critical capacity to discern, assess, and revise." Kaufman, In Face of Mystery, 193.
52. For a discussion of monological and dialogical ethical approaches, see Plumwood, Environmental Culture, 188–195.
53. Cf. ibid., 132–133.
54. Barad, Meeting the Universe Halfway, 182.
55. Feuerbach, Lectures on the Essence of Religion, 19.
56. Taylor, Erring, 25. In this book Taylor goes on to discuss the implications of the "death of the subject," "the end of history," and the "closure of the book" in ways that open up all these concepts to continuing creation in a way that I think theology ought to open up to the "many otherings" of creation. He writes, "When it no longer seems necessary to reduce manyness to oneness [God, self, history, text] and to translate the equivocal as univocal, it becomes possible to give up the struggle for mastery and to take 'eternal delight' in 'The enigmatical Beauty' of each beautiful enigma" (ibid. 176–177).
57. Keller, God and Power.
58. For a good critique of radical orthodoxy, see Ruether and Grau, Interpreting the Postmodern.
59. See, e.g., Keller and Schneider, Polydoxy; and Schneider, Beyond Monotheism.
60. Schneider, Beyond Monotheism, 162.

4. Destabilizing Identity
1. On the concept of "backgrounding," see Plumwood, Environmental Culture, 97–122.
2. Rubenstein, Strange Wonder, 7–12.
3. This is a process similar to what Hegel might describe as a synthesis between the ideal and material realms.
4. Mortimer-Sandilands and Erickson, Queer Ecologies, 3–4.
5. For an excellent film that draws from biblical scholars to debunk these eight passages, see Ky Dickens, Fish Out of Water (Chicago: Yellow Wing Productions, 2009).
6. Again, see Gudorf, "The Erosion of Gender Dimorphism."
7. Gayatri Spivak, "Can the Subaltern Speak?" in Nelson and Gross, Marxism and the Interpretation of Culture, 271–316.
8. Rubenstein, Strange Wonder, 74.
9. For a good analysis of the legal structures that continue to support straight supremacy, see Gabilando, "Asking the Straight Question."
10. Spade, Normal Life, 15–16.
11. See, e.g., Ruether, Christianity and the Making of the Modern Family.

12. Easton and Hardy, *The Ethical Slut*, 109.
13. On this process see, e.g., Suzanne Mrozik, "Materializations of Virtue: Buddhist Discourses on Bodies" in Armour and St. Ville, *Bodily Citations*, 15–47.
14. Anderlini-D'Onofrio, *Gaia and the New Politics of Love*, 124–125.
15. Val Plumwood, "Toward a Progressive Naturalism" in Thomas Heyd, *Recognizing the Autonomy of Nature*, 30.
16. Rubenstein, *Strange Wonder*, 112.
17. Much of this information on "intersexed" identities can be found at the Intersex Society of North America's Web site: http://www.isna.org/.
18. For a good history of trans issues, see Stryker, *Transgender History*.
19. See, e.g., Foucault, *History of Sexuality*.
20. Boellstorff, "Playing Back the Nation," 168. Cf. the queer theory classic study of "closet" ontology and epistemology: Sedgwick, *Epistemology of the Closet*.
21. Boellstorff, *The Gay Archipelago*, 202.
22. Ibid., 211.
23. Boellstorff, "Playing Back the Nation," 162.
24. Boellstorff, *The Gay Archipelago*, 44.
25. Scheibinger, *Nature's Body*.
26. Mortimer-Sandilands and Erickson, *Queer Ecologies*, 5.
27. This is what Latour argues throughout *The Politics of Nature*, for instance.
28. Boellstorff, *The Gay Archipelago*, 26.
29. Again, Richard Dawkins in *The God Delusion* still holds onto the Enlightenment mentality that truth is One, Single, and all must agree on it. This is nothing more than a hangover from a Greek and Christian transcendent monotheistic reality, which is the very thing that "modern science" loves to articulate itself against.
30. On this case, see Munro, "Caster Semenya"; and Hoad, "Run Caster Semenya, Run."
31. *The Telegraph*, October 12, 2009.
32. Boellstorff, "Playing Back the Nation," 161.
33. Lovejoy, "Reexamining Human Origins."
34. *Huffington Post*, October 1, 2009.
35. If this is the case, then we might agree with Marx that capitalism is a necessary evolutionary step along the way to communism.
36. From *Wall Street Journal*, October 3, 2009.
37. Goodchild, *Theology of Money*, 44.
38. Grau, *Of Divine Economy*, 57.

39. Such an identity formation takes inspiration from Rosemary Radford Ruether's "ecofeminist family ethic" in which families (here identities) become possible sites of prophetic redemption rather than solipsist sieves that reify all reality. See Ruether, *Christianity and the Making of the Modern Family*, 206–230.
40. Merchant, *The Death of Nature*.
41. Engels, *The Origin of the Family*.
42. This was the subject of my first book, *Theology, Creation, and Environmental Ethics*.
43. Plumwood, *Environmental Culture*.
44. See Boellstorff, *The Gay Archipelago*.
45. Haraway, *The Companion Species Manifesto*.
46. See Keller and Schneider, *Polydoxy*; and Althaus-Ried, *Indecent Theology*.
47. Mrozik, "Materializations of Virtue," 34.

5. The Emergence of Ecoreligious Identities

1. Robert Nelson also argues for these types of meaning-making systems as the two dominant forms of religion in the United States today. See Nelson, *The New Holy Wars*.
2. For a good discussion of the transformation of "technology," see Noble, *The Religion of Technology*.
3. Heidegger, *The Question Concerning Technology*, 3–35.
4. Kuhn, *The Structure of Scientific Revolutions*; and Whitehead, *Process and Reality*, 7.
5. Latour, *Reassembling the Social*.
6. Heidegger, *The Question Concerning Technology*, 152–153.
7. Deleuze, *The Fold*.
8. Keller, *Face of the Deep*, 169. Tehom is the depth or watery chaos at the beginning of creation in the Genesis story. Keller articulates it as the source of creativity and possibilities.
9. Haraway, *Simians, Cyborgs and Women*.
10. This is what Bruno Latour suggests as well in his understanding of that which gets left out of any collective. It is the excluded that comes back to challenge the boundaries of the collective, open it up to otherness, and hence reorganize into a new collective, ad infinitum. Latour, *Politics of Nature*.
11. Derrida, *Specters of Marx*, 161.
12. Spivak, *Death of a Discipline*, 71–102.
13. Heise, *Sense of Place and Sense of Planet*, 17–67.
14. Julie Gold, "From a Distance" (1987).
15. Brennan, *Globalization and Its Terrors*.
16. Taylor, *Dark Green Religion*, 84–85.

17. Brennan, *Globalization and Its Terrors*, 66–95.
18. Bauman, *Globalization*.
19. You can complete your own "ecological footprint" by visiting the Web site of Redefining Progress: http://www.ecologicalfootprint.org/.
20. Spivak, *Death of a Discipline*, 71–102.
21. Again see Keller and Kearns, *Ecospirit*, 1–20.
22. See Haraway, "Situated Knowledges."
23. Tweed, *Crossing and Dwelling*, 81.
24. Plumwood, *Environmental Culture*, 50–61.
25. See Merchant's analysis of this in *Reinventing Eden*.
26. Bauman, *Liquid Modernity*, 164.
27. Bruno Latour, "Thou Shalt Not Freeze-Frame" in Proctor, *Science, Religion and the Human Experience*.
28. Deacon, *Incomplete Nature*, 539.
29. Those familiar with process thought will recognize this type of subjectivity. Though heavily influenced by process, I do stray from the metaphysical system for multiple reasons that are beyond the scope of this chapter. I have discussed these elsewhere: Bauman, *Theology, Creation, and Environmental Ethics*, 162–163.
30. Morton, *The Ecological Thought*, 15.
31. Ibid.
32. Haraway, *The Companion Species Manifesto*.
33. Deleuze and Guattari, *A Thousand Plateaus*, 149–166.
34. See, e.g., Haag and Bauman, "De/Constructing Transcendence."
35. McIntosh, "White Privilege and Male Privilege."
36. Of the many good sources on this topic, see Marable, Steinberg, and Middlemass, *Racializing Justice, Disenfranchising Lives*.
37. See, e.g., Dieter, *The Death Penalty in Black and White*.
38. On the issue of "per/versions," see Althaus-Reid, *Indecent Theology*.
39. "'Subjection' signifies the process of becoming subordinated by power as well as the process of becoming a subject." Butler, *The Psychic Life of Power*, 2ff.
40. Claudia Schippert, "Turning On/To Ethics" in Armour and St. Ville, *Bodily Citations*, 157–176.

6. Developing Planetary Environmental Ethics

1. Heise, *Sense of Place, Sense of Planet*, 42.
2. Karen Barad, in *Meeting the Universe Halfway*, also uses the concept of epistemontology. I derived it from thinking through Gordon Kaufman's understanding of biohistory and Donna Haraway's understanding of nature-cultures independent

of Barad's work. It was thus a pleasant surprise to find someone else using the same language. See, e.g., Bauman, "The Eco-Ontology of Social/ist Eco-Feminist Thought."
3. Soper, Ryle, and Thomas, *The Politics and Pleasure of Consuming Differently*, 3.
4. Plumwood, "Shadow Places and the Politics of Dwelling," 140.
5. Bergmann and Sager, *The Ethics of Mobilities*. Cf. Plumwood, *Environmental Culture*, 218ff.
6. Sigurd Bergmann, "The Beauty of Speed or the Discovery of Slowness" in Bergmann and Sager, *The Ethics of Mobilities*, 17. This volume contains some constructive notions of identities of mobility and other chapters that are critical of mobility and long for the return to place. See, e.g., Juhani Pallasmaa, "Existential Homelessness—Placelessness and Nostalgia in the Age of Mobility" ibid., 143–156.
7. Bergmann, "The Beauty of Speed," 23.
8. Peter Nynas, "From Sacred Place to an Existential Dimension of Mobility," in Bergmann and Sager, *The Ethics of Mobilities*, 173.
9. Brennan, *Exhausting Modernity*, 126. I want to thank Marty Reineke for pointing me toward Teresa Brennan's work. Cf. Michael Northcott, "The Desire for Speed and the Rhythm of the Earth," in Bergmann and Sager, *The Ethics of Mobilities*, 215–232.
10. I use globalatinization in the sense that Derrida uses it to connote the unequal power dynamics in globalization. He uses globalatinization precisely to connote the ways in which Western ideals and processes are being forced upon the entire planet. See, e.g., Derrida and Vattimo, *Religion*.
11. Juliet Solomon, "Happiness and the Consumption of Mobility," in Soper, Ryle, and Thomas, *The Politics and Pleasure of Consuming Differently*, 169.
12. I am drawing on some of Jurgen Moltmann's (via Ernst Bloch) distinctions of time here. By future-present, I simply mean a future that is closed off by thinking in the present. In contrast, a future-future is radically open for genuine, new, emergent possibilities that can never be contained by present thinking.
13. Keller, *Apocalypse Now and Then*.
14. Žižek, *In Defense of Lost Causes*, 439–440.
15. Plumwood, "Shadow Places and the Politics of Dwelling," 141.
16. Brennan, *Exhausting Modernity*, 7–8.
17. Ibid., 34.
18. I use one-fifth world to refer to "developing," or "first" world nations. This designation is one of the few that does not have a qualitative association with it. Further, it highlights the fact that those living in modern Western, technologized, consumerist settings make up a minority of the world's population.
19. Bauman, *Globalization*.
20. Ibid., 123.

21. Ibid., 129.
22. See, e.g., Keller, *Face of the Deep*; and Bauman, *Theology, Creation, and Environmental Ethics*.
23. See, e.g., Rick Del Vecchio, "Cal Sees BP Deal as Landmark: Research Could Lead More Quickly to Making Alternative Fuel a Reality, at SFGate.com, February 2, 2007.
24. Herzogenrath, *Deleuze/Guattari and Ecology*, 55.
25. Morton, *Ecology Without Nature*; Latour, *Politics of Nature*.
26. Tucker, *Worldly Wonder*, 31 (my emphasis).
27. One implication of this is that I do not think any religion or meaning-making practice is "ready-made" to address ecological ills. There may be similarities between codependent arising (in Buddhism), but this was not a foreshadow of the modern science of ecology. Hence all religions, even those with very promising ecological insights and ideas, must work to renegotiate meaning in light of contemporary ecological ills.
28. Keller and Kearns, *EcoSpirit*.
29. My current favorite example is the drive for marriage in the GLBTQ community. Isn't this (to a certain extent) just placing GLBTQ loving relations within the structure of heteronormative realities? Might a better solution be found in fighting for universal healthcare (rather than the same rights as straights), ending tax favors for a certain group of people over all others (rather than joining in on those tax breaks), or fighting to end the industrial military complex (rather than asking for open representation within it)? I thank Steven Blevins and Jose Gabilondo at Florida International University for extensive conversations around these issues.
30. Brennan, *Exhausting Modernity*, 129.
31. Catherine Keller and Laurel Kearns, "Introduction," in Keller and Kearns, *EcoSpirit*, 4.
32. Bruno Latour, "Thou Shalt Not Freeze-Frame" in Proctor, *Science, Religion, and the Human Experience*, 36.
33. Tweed, *Crossing and Dwelling*, 54.
34. Ibid., 112–113.
35. Spivak, *The Death of a Discipline*, chapter 3. See also the recent volume edited by Moore and Rivera, *Planetary Loves*.
36. Heise, *Sense of Place and Sense of Planet*, 17–67.
37. Tweed, *Crossing and Dwelling*, 81.
38. Žižek, *In Defense of Lost Causes*, 445.
39. Lovejoy, "Reexamining Human Origins in Light of Ardipithecus Ramidus"; Frans de Wall, "Our Kinder, Gentler Ancestors," *Wall Street Journal*, October 3, 2009.

40. The late Val Plumwood spoke of this as the "ontological vegan" versus the "ethical vegetarian" problem. The ontological vegan operates out of the logic of domination and imposes her view across the face of the globe. The ethical vegetarian operates on a more dialogical basis, negotiating cultural and religious differences in determining when it is or is not all right to hunt and/or use animal products. See Plumwood, *Environmental Culture*, 143–166.
41. I assume that Locke must have been aware of this implication as he placed strict limits on private property to that of one's own labor. Of course, this limitation has not panned out.
42. See, e.g., Bauman, *Theology, Creation, and Environmental Ethics*, 67–87.
43. Most of these critiques came (infamously) from his critiques of feminism in Berry, "Feminism, the Body, and the Machine."
44. There is a lot of literature on perception and the process of transduction and transmission in processing information. For a good textbook introduction, see Goldstein, *Introduction to Perception*, 3–22.
45. Latour, "Thou Shalt Not Freeze-Frame," 14. Similarly, Luis Dupre notes: "Which ideas triumph often depends less on inner logic than on the rhetorical power and ideological simplicity of those who defend them." Dupre, *Passage to Modernity*, 58.
46. Shellenberger and Nordhaus, "The Death of Environmentalism." A group of environmental justice advocates wrote a very critical response indicating, rightly, that Shellenberger and Nordhaus had overlooked the environmental justice movement: Gelobter, Dorsey, Fields et al., *The Soul of Environmentalism*.
47. Terrence Deacon, "Emergence: The Hole at the Wheel's Hub" in Clayton and Davies, *The Re-Emergence of Emergence*, 111–150.
48. There is a lot of literature on the importance of "play" and on the importance of meditative practices for promoting environmental sustainability. See, e.g., Tucker and Grim, *Religions of the World and Ecology*. See also several of the chapters in Bergmann and Sager, *The Ethics of Mobilities*.
49. Solomon, "Happiness and the Consumption of Mobility," 163.

7. Challenging Human Exceptionalism

1. Žižek,, *The Fragile Absolute*, 95.
2. Deleuze and Guattari, *What Is Philosophy*, 212.
3. Colebrook, *Gilles Deleuze*, 129. Donna Haraway's cyborg ontology and concept of companion species is also relevant here.
4. Rosi Braidotti has written a very helpful book on the notion of nomadic identities: *Transpositions*.
5. Kung, *Global Responsibility*.

6. See, e.g., Sabrina Tonutti, "Anthropocentrism and the Definition of 'Culture' as a Marker of the Human/Animal Divide," in Rob Boddice, Anthropocentrism, 183–202.
7. Plumwood, Environmental Culture, 159–166.
8. Val Plumwood, "Being Prey," in O'Reilly, O'Reilly, and Sterling, The Ultimate Journey, 128–148.
9. Plumwood, Environmental Culture, 156–158.
10. This is what Karen Barad is attempting in her book Meeting the Universe Halfway, as discussed in more detail in previous chapters.
11. Haraway, When Species Meet, 66–67.
12. Turing, "Computing Machinery and Intelligence." Turing's test suggested that when a human observing responses in a blind experiment could not tell the difference between a computer-generated response and a human response we should treat the responses equally. As such, we might here say if it looks and sounds like history, why not treat it as history?
13. Plumwood, Environmental Culture, 132–138.
14. Barad, Meeting the Universe Halfway, 137.
15. Ibid., 116.
16. Ibid., 140.
17. Spencer, Gay and Gaia. It's a reference to the entire work, so the note can be removed and it can remain in the bibliography.
18. Latour, We Have Never Been Modern.
19. Heidegger, The Question Concerning Technology, 3–35.
20. Those in linguistic studies have spilled much ink over these issues, and, as I am not a linguist, I will leave this brief digression on its own.
21. Pinker, The Better Angels of Our Nature. I thank Catherine Keller for helping me to think through the inherent "antiprogress" mentality in much post-thinking today.
22. Keller, God and Power, 150.
23. See Lovelock, Gaia.
24. Deleuze and Guattari, A Thousand Plateaus, 164–165.
25. Bauman, Creation, Theology, and Environmental Ethics. This work is in close dialogue with Catherine Keller's critique of ex nihilo in her Face of the Deep.
26. David Byrne, Brian Eno, Chris Frantz, Jerry Harrison, Tina Weymouth, "Once in a Lifetime," on Remain in the Light (New York: Sigma Sound Studios, 1981).
27. Arendt, Eichmann in Jerusalem.
28. Nixon, Slow Violence and the Environmentalism of the Poor.
29. Abram, Becoming Animal, 27.
30. Kaufman, The Theological Imagination.
31. Keller, Face of the Deep, 155ff.

GLOSSARY

Please note that the following glossary uses operational definitions of the terms listed. In other words, they are meant to help you understand the words as they are used in these chapters and as the author understands them to be used by theorists discussed throughout the book. Each of these words has many different connotations and contexts, so none of the following glossary entries is exhaustively defined. My suggestion to the reader that wants to find out more about these words is to follow the notes in the text, find out where these words are being used, and look up some references to them. Another helpful source with much more information is the online *Stanford Encyclopedia of Philosophy*: http://plato.stanford.edu/.

ABJECT (ABJECTION). The abject is that which is left out of identity construction or normative statements. In other words, it is that which is queer and does not fit a categorization. This is used in reference to identity construction in queer theory.
AESTHETIC. Aesthetics is the study of beauty. Aesthetics provides a source for doing ethics that asks what types of worlds we want to help to create (questions of beauty) rather than questions about what is good or evil necessarily.
AGNOSTIC (AGNOSTICISM). Agnosticism is a valid position in many philosophies and epistemologies that suggests we can never get beyond our human perspective to know reality in full. Historically, in theological terms, it has been a term that navigates between *theism* and *atheism*.

AGRIPPA'S TRILEMMA. Agrippa (Ancient Greece) discussed three options for justifying any truth claim: foundationalism (a statement that founds a claim in a reality such as God or Platonic Forms); circularity (A relies on B, which relies on C, which then relies on A again); or infinite regress (there is no way to secure our truth claims, and one can continually ask "why" ad infinitum).

ANARCHISM (ANARCHY). Anarchy is the negation of *arche* or origins. In other words, it is a position that suggests there is no ultimate (original) reality by which we can judge our ethical claims.

ANACHRONISTIC (ANACHRONISM). A claim is anachronistic or an anachronism when it imposes an idea from the present context onto the past. For instance, sexual orientation did not exist as we know it in today's Western world, therefore suggesting that the Bible discussed homosexuality is an anachronism.

ANEKANTA (NONABSOLUTISM). This is a term that comes out of Jain philosophy. It suggests that there is no perspective (save the enlightened beings) that can ever exhaust all reality. Thus, our language is limited and never objective.

ANIMISTIC (ANIMISM). Animism suggests that the entire physical world is somehow alive or animated. Henri Bergson's *élan vital* is a good example, as is the idea found in Jainism that all of life is ensouled. Many indigenous traditions, including Shinto in Japan, also view the world in some type of animistic terms.

ANTHROPIC PRINCIPLE. The anthropic principle is a theory that suggests the universe in which we live was fine-tuned to produce human life. In other words, we were somehow meant to emerge in the universe from the big bang some 13.7 billion years ago. Some use cosmological constants such as gravitational pull and the rate of expansion to suggest that our universe is very unlikely due to chance and therefore had to be this way. Others loosely use the anthropic principle to still justify some sort of belief in a theistic God. Still others challenge the anthropic principle suggesting that we might actually live in a multiverse of which our known universe is one of billions.

ANTHROPOCENTRISM. Anthropocentrism is thinking that is human (*anthropos*) centered (*centrism*). It can involve the anthropic principle, but it always suggests that humans are the center of meaning and value and are thus somehow ethically above other life-forms on the planet. It can also be used to suggest that the rest of the natural world is merely a resource for human ends. The ideas of dominion and stewardship have both been criticized as being anthropocentric in the field of environmental ethics.

APOCALYPTIC DISCOURSE (APOCALYPTICISM). An apocalypse is the unveiling of reality. It is a particularly important type of literature for prophetic Scriptures. The most common story people associate with the apocalyptic in the West is the book of Revelation. Apocalypticism is known for its gloom and doom critiques of reality as we know it. As such, it has been a popular form of discourse for environmentalists in talking about ecological problems such as global climate change.

APOPHATIC (APOPHASIS). Apophatic thinking is a theological method that suggests we can never exhaust the ultimate reality of God. In other words, whatever that ultimate reality is, our human ability to know that God is limited. This means that when we talk about god, God or, in the context of this book, ultimate reality, we have to admit some uncertainty at the edges of our knowledge. It is this apophasis that works against idolatry or the idea that we can be like god and know reality in full.

A PRIORI. A priori statements are foundational statements in epistemological claims. In other words, they are the rock bottom truths that one argues for in knowledge claims. The philosopher Immanuel Kant argues for such "categorical imperatives" in knowledge claims; Descartes argues for them in his "clear and distinct thoughts," and Plato's Forms are still another version. The point of such claims is that they have nothing to do with context; rather, they are the same for all times and all places; they are just arrived at or intuited.

ARBOREAL THINKING. The French philosophers Deleuze and Guattari refer to foundational types of claims as arboreal. That is, knowledge has a root or origin that can be discovered and from which all other knowledge flows. The same can be said in terms of identities: there is some root, such as the individual soul, that secures a given identity. Such arboreal thinking also works its way into reductive methods of understanding in the natural sciences: "it's all in the genes" or "it's all in the neurons" would be two such examples. They contrast this way of understanding knowledge to rhizomatic thinking.

ARCHIMEDEAN STANDPOINT. From the Greek story of Archimedes, this is the idea that we can somehow find an objective point or space removed from the world from which we can see all reality clearly. This is the god's eye view of the world that is refuted by all forms of postmodern thinking.

ARCHIPELAGIC SELF. Anthropologist Tom Boelstorff argues that Indonesian identity is different from Western identities. Whereas Western identities are founded upon a single, enduring, essential identity (perhaps based on the idea of being made in the image of a monotheistic god), Indonesian identities are founded upon multiplicities. Like a series of islands, an identity can be many different things in different places. In this book I use this as a model for understanding our subjectivities as multiple.

ASSEMBLAGES. This is yet another metaphor for understanding our subjectivities, as well as all entities/processes/organisms, as always already multiple. Drawing from the works of Deleuze and Guattari, the idea is that concepts such as tree, self, dog, or species are always making a superficial cut in how they are defined. That is, such understandings cut living, evolving entities off from the dynamic processes, histories, and flows of information that make up an entity at a given time. Assemblages is a way to name that always already multiple entity.

AUTOPOEISIS. Autopoesis, or self-organizing/self-creating, is a way of describing how new things come into being, especially in the context of emergence theory. It is also used as a term to prevent the reification or closure of life into human concepts. Broadly, it can be seen as naming the ongoing creative process of nature becoming with no definite goal or end in sight.

CATAPHATIC THEOLOGY. Cataphatic or positive theology is a positive, affirming way of talking about the divine. In this model we can know and speak of God or ultimate reality. In other words, there is a reality that our language and concepts can reach, know, and represent. From this perspective, knowledge can be closer to or further from ultimate truth.

CAUSALITY (MATERIAL, EFFICIENT, FORMAL, FINAL). Normally when we think of causality, we think of efficient causality. However, Aristotle, among many others now including the emergentists, reminds us that there are at least four types of causality. Material causality is based upon the substance that a thing is made of: A bird without feathers cannot fly, thus the material is important. Efficient causality would be the bird's muscles moving to contract and expand the wings. The formal causality would include the aerodynamics of flight, and the final causality might be that the bird was trying to go hunting for food to bring back to its nest (or digesting and defecating seeds that will eventually become new plant growth). All these causes are relevant, but, since the time of the scientific revolution, Westerners have mostly been concerned with efficient causality (which some suggest is a real flaw with modern ways of understanding the world).

CISGENDERED. This simply means that one was born with a sex and gender that match the norms of what that sex/gender are in a given society and with which s/he identifies. In other words, rather than suggesting that transgender is somehow abnormal, the word *cisgender* suggests that being born to have one's sex/gender match that with which s/he identifies with is equally normal as being born transgendered; they are just different ways of being born.

CONSTRUCTIVISM (CONSTRUCTED/CONSTRUCTION). The idea of constructivism is that our knowledge and our identities are completely constructed. This is drawn in sharp contrast to essentialist or foundational thinking in which identities and knowledge can be true and real regardless of context. Constructivists most often argue that all reality is shaped by language or culture and that these contexts can never be escaped. They are most often associated with relativism (though I argue that this is not always an accurate characterization).

CONTEXTUALISM (CONTEXTUALITY). Contextuality is a position that navigates between universal, essential, objective-style thinking and identity formation and complete constructivist/relative positions. Contextualism would argue that all reality is a coconstruction (between animals/humans, biology/histories, nature/cultures) and that such a position pays attention to bodies and realities more than an objective/universal or constructive/relative position.

CONVIVINCIA. The *convivencia*, or "living together," was a period from about the eighth and fifteenth century in southern Spain when Jews, Christians, and Muslims lived together in relative peace sharing ideas, art, and culture. Though scholars debate just how peaceful this period was, this was a time when all three cultures could be found together in one place. One famous institute built during this period was the Alhambra in Andalusia, and one famous city of this period was Cordova (also in Andalusia).

COPERNICAN REVOLUTION (SEE ALSO GALILEAN REVOLUTION). The Copernican Revolution is usually thought of as the point at which the modern scientific revolution began. It was based upon Copernicus's mathematical arguments for a heliocentric universe, which argued that the earth was not at the center of the universe. Galileo later confirmed this observation with his telescope. The idea that Copernicus, Galileo, or even the West is responsible for the rise of modern science in total denies the contributions to science made by Muslims during the Golden Age of Islam and by Chinese and Indic cultures.

CREATIO EX NIHILO. This is the idea, formulated by Irenaeus, Tertullian, and Athanasius in the second to third century of the common era, that God (the Christian God) created the entire universe out of nothing. Such an assertion went against the Greek idea that "nothing from nothing is nothing" and that therefore matter and forms must be somehow coeternal. Furthermore, it went against other ideas of the eternal return or cyclical time in the Ancient Near East. Finally, it provided a strong rhetorical truth claim once Christianity came to be the religion of empire: in other words, my God created everything out of nothing and therefore must be the true God.

DADAISM. Dadaism is an art movement of the early to mid twentieth century that was steeped in the absurd and the philosophy of anarchy. It is a response to the modern obsession with form and function, and at heart it is a recognition that creativity cannot be controlled (so to speak). To get a flavor of Dadaism, one can Google the Dada manifesto and read one or both of the Dada manifestos written around 1918.

DARK AGES (MEDIEVAL PERIOD). The dark ages or the medieval period are generally considered by Western historical accounts as the time from the fall of Rome in the fifth century until the Renaissance in the twelfth century, when philosophers and theologians in the Latin West began reading and studying the ancient Greeks once again. This historical narrative completely writes over the Golden Age of Islam from the sixth to the twelfth centuries, during which much writing in natural philosophy, cosmology, optics, and medicine was done in Arabic.

DEATH OF GOD. The death of God is attributed first and foremost to Nietzsche's famous statement that God is dead in his book, *Thus Spake Zarathustra*. What Nietzsche argued was that the metaphysical omni-God that holds all of reality into being was dead. Subsequent philosophers and theologians have taken up this death of God move as a place from which to critique theology and philosophy that rely on substance metaphysics and ontology or the study of being also referred to by Heidegger as ontotheology. The idea is that, whatever god might be, god is of more ethical and aesthetic importance than an actual being that secures reality, knowledge, and truth in some sort of foundational or teleological way.

DECONSTRUCTIONISM (DECONSTRUCTION). Deconstructionism is a method that draws deeply from Nietzsche's death of God and more broadly from the idea that concepts, words, and even norms are created over time. In other words, there is nothing God given or natural. What we come to see as God given or natural has been constructed over time. Every time we associate our concepts with reality or truth, there is something left out, something differed (see the abject), and this remainder haunts the definition, concept, or term that we are using. The most prominent deconstructionist is Jacques Derrida.

DEPENDENT COARISING (ALSO PRATĪTYASAMUTPĀDA OR DEPENDENT COARISING). This is the idea that everything is connected. One cannot separate things out one from another, nor can we separate ourselves from human and earth others (present, past, and future). This is one of the major tenets of Buddhism and is quite different from the individualistic identities and substance-based metaphysics of, say, the Abrahamic faiths. In this perspective there is no self or other that is separate from other selves and others; thus change, interaction, and interrelatedness are what mark the true nature of reality.

DETERRITORIALIZE/RETERRITORIALIZE. This is a method that follows the work of French philosophers Gilles Deleuze and Felix Guattari. The point of thinking, from this perspective, is to be iconoclastic; that is, we should never settle into our language and concepts as if they capture reality. Rather, we ought to pay attention to what is left out. That which is left out deterritorializes concepts, identities, and knowledge. We will always reterritorialize, but this process of deterritorializing, which helps us to pay close attention to the "remainder" and reterritorializing is endless. There is never a point at which we arrive at the truth or some sort of stasis.

DIFFERANCE. This refers to that which is left out of a concept or that which is deferred or different from any category that we might come up with. This is what allows the process of thinking and ethical deliberation about how we ought to become to continue. It is, again, a process of defining, deferring, and then redefining based upon that which challenges the concept or truth claim or identity that has been decided upon.

DOMINOLOGY. Dominology describes the process by which one's own ideas or ways of being in the world are imposed on the rest of the world. This is characteristic of colonial ways of thinking including those found in the process of Western globalization, which assumes, very simplistically, "my way is the best way." Metaphors of civilizing, Christianizing, developing, progress, and modernization can all be examples of dominology.

ECONOMIC GLOBALIZATION. Economic globalization is the spread of free-market style capitalism and industrialization over the face of the entire planet. Many would argue that, after the fall of communism in 1989, There IS No Alternative (TINA) or that capitalism is the only way to economically organize the world. Structures such as the World Bank, the International Monetary Fund, and other "Bretton Woods" institutions enforce this idea of the inevitability of capitalism. Many peoples from all over the world would argue that such an imposition is not leading to development but rather impoverishment of local places and that it is leading to the increasing gap between the very wealthy and the majority of people that are poor. Also, many would argue that economic globalization is the root cause of the environmental crises we all face today because it does not account for the rest of the natural world outside of its use for human beings.

ECOTONE. An ecotone is the area between two different ecosystems that works as an in-between space that allows for the transfer of energy and materials between ecosystems. Such a transfer is necessary for the ongoing evolution and change of any ecosystem; thus ecotones are crucial for keeping ecosystems open to the process of change over time. In this text it corresponds to the interstitial space of identity formation.

ÉLAN VITAL. The *élan vital* or life force is a concept developed primarily by Henri Bergson to suggest that all life is alive, agential, and moving toward a (perhaps unknown) future. This strictly contrasts with the Western modern scientific division between humans (as alive and agential) and the rest of the natural world (as dead matter). Many contemporary emergent theorists draw on and modify Bergson's notion of the *élan vital*.

EMERGENTISTS (EMERGENCE THEORY). Emergence theory is important for this text in two primary ways. Emergence seeks to navigate a way between idealism (or top-down) thinking and reductive materialism (or bottom-up thinking). In this sense emergence is a system of thought that seeks to understand how new things come to being in the evolutionary process that cannot be accounted for by reducing things to lower (material) levels or greater (divine) intervention. Second, emergence theory seeks to navigate between radical newness (from nowhere) and simply a reordering and reshuffling of the same. In other words, human brains rely on the material, neural, chemical, and genetic levels, but consciousness cannot be reduced to these levels. Consciousness is an emergent phenomenon in the process of evolution.

ENFOLDING (THE FOLD). Rather than understanding existence as a climb upward toward some transcendent horizon or as capable of being deduced from any ultimate standpoint, enfolding suggests a radically immanent way of understanding how material and physical reality are always intertwined with ideas, imagination, and meaning. In other words, this is a metaphor for thinking about how to bring many different levels of reality (material, imaginative, psychological, religious, genetic, chemical, energetic, and technological) together into a single idea rather than arguing for some sort of dualism between mind/brain, thought/matter. Further, rather than understanding one's experiential horizons as ultimate, experiential horizons should be seen as folds in space-time and as part of others (past, present, and future) horizons.

ENTROPY. Entropy is, in physical and scientific terms, the second law of thermodynamics. This law suggests that in closed systems, as energy is exchanged, entities tend toward lower and lower levels of energy or equilibrium. Cosmologists are trying to understand how the increase in complexity through evolution (and gravity in general) exist along with a trend toward further and further dissipation of things (or cosmic expansion, entropy). In cosmology, theories that attempt to explain how gravity and the tendency toward complexity can exist at the same time as a trend toward cosmic expansion, and energy dissolution, are known as theories of everything (or, more tongue-in-cheek, as a TOE).

EPISTEMIC ANTHROPOCENTRISM (VERSUS ONTOLOGICAL ANTHROPOCENTRISM). The distinction between epistemic or knowledge-based anthropocentrism and ontological or an anthropocentrism that understands humans as metaphysically at the center of all life in the universe, is one that is made by Australian environmental philosopher Val Plumwood. She argues that, because we are contextual beings, we cannot help but be epistemically anthropocentric—that is, we only know the world from our human perspectives. However, there is no reason why we must therefore be ontologically or ethically anthropocentric—that is, there is no reason why we must think of human beings as the culmination of the expansion of the cosmos or of humans as the center of all ethical concern. We are one, located species, but this does not mean we can treat all other species as if they exist for the use and concern of human beings.

EPISTEM-ONTOLOGY. This term is grounded in the work of thinkers such as Donna Haraway, Deleuze and Guattari, and the pragmatists. The idea is that there is never a separation between knowing (or what we can know) and being (or reality as it exists). This means that what we will always influence the way in which the world is made. Moving beyond mere humanity, the idea is that there is no clear direction toward which life should evolve, but all organisms determine the direction in which life moves; to be sure teleodynamics, or goal-directed processes, emerge, but this does not mean they were inherent to the structure of the universe at the time of the big bang. Thus, there is no clean cut between human knowing and the way in which the world becomes. Knowledge (whether genetic repetition or chemical reactions or consciousness) shapes the way that life evolves, but not in any sort of direction that was predetermined for all times and all places.

ESSENTIALISM. Essential thinking is the idea that there is a certain way that things, peoples, animals, and reality are supposed to be. In other words, we are born into a world that is the way things are and are judged as meeting that reality in terms of how much we conform to that reality. Aristotelian teleology is one classic example of essentialism.

EVENT-BASED ONTOLOGY. An event-based ontology or metaphysic (such as found in the process thought of Alfred North Whitehead and the philosophy of Alain Badiou) argues that events, processes, and change are the true nature of reality rather than any substance, essence, or unchanging form. From this perspective, as the Greek pre-Socratic thinker Heraclitus famously quipped, "you can never step in the same river twice."

FOUNDATIONALISM (THE FOUNDATIONAL FANTASY). Foundationalism is the idea that reality is somehow secured by an eternal essence, form, or being. Such foundations are true for all times and all places throughout the 13.7 billion years process of cosmic expansion. The foundational fantasy is a term that post-foundational thinkers use to suggest that any such universal claim is a fantasy. It only seems universal, these post thinkers might argue, because the historical construction of the foundational truth claim has been hidden or lost. The method of genealogy in the works of Nietzsche and Foucault is an attempt to uncover the hidden historical constructions of such foundations.

FUNDAMENTALISM. Fundamentalist thinking, not to be confused with evangelical thinking, is associated with a collapse of knowledge and reality in such a way that one's belief or truth claim is unshakably the only truth or the only right belief. The charge of fundamentalism is often made against religious peoples (such as Christians/Muslims that believe the Bible/Quran is the inerrant word of God/Allah), but it can also be used to describe strict reductive materialists who argue that everything can be explained by genes, neurons, and/or chemicals.

GALILEAN REVOLUTION (SEE ALSO COPERNICAN REVOLUTION). The Copernican revolution is usually thought of as the point at which the modern scientific revolution began. It was based upon Copernicus's mathematical arguments for a heliocentric universe, which argued that the earth was not at the center of the universe. Galileo later confirmed this observation with his telescope. The idea that Copernicus, Galileo, or even the West is responsible for the rise of modern science in total denies the contributions to science made by Muslims during the Golden Age of Islam and by Chinese and Indic cultures.

GENDER DIMORPHISM (SEX DIMORPHISM). Gender and sex dimorphism are ideas that there are only two genders/sexes in the world: male and female. One is either male or female and there is no in between. Queer theory, gender studies, as well as scientific studies of biology challenge this idea. Furthermore, this idea is challenged by the fact that third and fourth and in-between genders/sexes exist in many different cultures.

GLOBALATINIZATION. *Globalatinization* is a term coined by Jacques Derrida to describe the unequal process of globalization. In other words, what is happening in globalization is not an equal mixing of cultures but much more the imposition of Western culture and economics over the face of the globe. Globalatinization captures this unequal process better than just globalization.

GLOCAL. The *glocal* is a term used in sociology (primarily but now more widely across the academy) to suggest that every local place on the globe is always already caught up in global flows of information, energy, materials, histories, and cultures. In other words, there is no local that is not already part of global networks.

GOLDEN AGE OF ISLAM. The Golden Age of Islam was a period from about the sixth to twelfth centuries of the common era during which Islamic civilizations flourished from contemporary Portugal and North Africa all the way to the edge of China. This was the period called the Dark Ages by the Christian West, but during this time all of the Ancient Greek thinkers were translated into Arabic and there were many advances in optics, medicine, chemistry, and other sciences.

GREEN REVOLUTION. Despite its name, the green revolution was far from environmentally beneficial. This was a process and idea developed by the United Nations after World War II in an attempt to stamp out poverty. The idea was to export industrial-style agriculture of the West to all parts of the globe. Such technologically intensive agriculture is quite expensive, so while yields of crops in places like India and South America increased under this plan, actual poverty increased because farming become more and more consolidated under fewer and fewer landowners. As crop yields increased, this sent prices for crops down, forcing smaller farmers out of business and leading to land consolidation by those who could afford the new technologies. This is a perfect example of how good ideas can go awry in a global context.

HELIOCENTRIC (HELIOCENTRISM). This is the Copernican and Galilean cosmology that displaced the geocentric understanding of the cosmos in the fifteenth century. These cosmologies argued that the sun was at the center of the universe: heliocentrism. This proved to be a serious problem for religious thinkers who had assumed that humans were at the center of the universe or the center of God's universal plan for salvation. On the one hand, heliocentrism displaced the importance/centrality of the earth. On the other hand, the Ptolemaic geocentric universe also ranked the four elements fire, air, earth, and water in a hierarchy of value. The earth was at the center and lowest place because it was the heaviest/lowest element. Replacing the earth with the sun also challenged the reigning Aristotelian notions of the makeup of the cosmos.

HETERONORMATIVITY. This is the assumption that the norm is heterosexuality. Thus legal, educational, political, medical, religious, and other systems are set up to assume heterosexuality. One assumes that a person is heterosexual until she says otherwise. This is usually a covert form of systemic discrimination against those who identify as LGBTQ.

HETEROSEXISM. Heterosexism, like sexism, racism, and other discriminatory isms, is the belief that heterosexuality is the superior, normal, and correct expression of sexuality. Thus this is a more overt and intentional form of discrimination against individuals who identify as LGBTQ.

HOLISM. Holism is a way of understanding reality that is in contradistinction to dualism and atomism. In other words, it suggests that all reality is interrelated. From a post perspective, wholism is problematic because it suggests that there is a whole/unity into which all the diverse components of reality can fit. Thus, in the attempt to overcome the problems of colonial thought (that nature is dead reality and that there is a dominant truth that all others can fit into), wholism ends up ordering life into a vision of centrism whereby all things are related to the whole. This turns out to be centrism all over again, because the order of the whole depends on the one(s) defining the whole.

HUMAN EXCEPTIONALISM. Human exceptionalism is the persistent, yet often underground assumption that human begins are somehow exceptions to the rest of the natural world: whereas all other things are material, humans are spiritual; whereas all other species evolve and will eventually go extinct or evolve into something new, humans are somehow different, and there is something ethically superior to being human. This is at the heart of ethical anthropocentrism and suggests that we are not of this earth.

HYBRID (IDENTITIES)/HYBRIDITY. Hybridity and hybrid identities are terms that suggest we, as human begins, are mixtures of nature/culture, material/ideal, and organic/machine (Donna Haraway), but also that there is no pure identity, tradition, or concept. Rather than having pure identities or essences, everything is in the mix, and it is from this place of hybridity that we should begin to think about reality.

HYPEROBJECT/HYPERSUBJECT. Hyperobjectivty and hypersubjectivity are both terms that highlight the mixed nature of our entities, organisms, identities, and concepts. Furthermore, these terms suggest that there is no reason why we ought to analyze and think about life at, say, the individual human level, when every individuality is made up of multiple times, places, minerals, bacteria, technologies, and histories. Each of these levels and each entity or organism is part of a larger evolving process: evolutionary, sociological, and even cosmological. The unit of the individual has been particularly strong in modern Western modes of thought, but it is not necessarily the best unit of analysis or the only unit of analysis.

ICONOCLASM. Iconoclasm is a method for breaking reified images. In other words, it is a method within philosophy, and religious studies especially, that works to deconstruct concepts that have come to be assumed as normal and/or natural. It is also a method for pointing out the injustices and inaccuracies in accepted ways of thinking and acting toward human and earth others.

IDEALISM (HEGELIAN). Idealism, especially in its Hegelian sense, is a form of thinking about the world that suggests that the really real is something that transcends the everyday material realities we experience. Salvation histories that suggest everything will be made right in the end are forms of idealism as well. This stands in direct opposition to Marxist materialism, though I along with others argue that idealism and materialism are mirror images of one another.

IMMANENCE. An immanent understanding of the world argues that whatever consciousness, spirituality, or divinity might be, they must exist in the same realm as the everyday world of material existence. In other words, there is not a transcendent space known as heaven or salvation that is in direct dualistic opposition with the material world, nor is the mind or the soul something that is in a different space from that of the brain or the body. Speculative realism, secular theologies, radical materialism, and event-based ontology are just a few names contemporary methods that work with immanent understandings of reality.

INDRA'S NET. This is a Buddhist metaphor to describe the interrelated and dependent coarising nature of the universe. In other words, everything exists in interrelatedness like a spider's web so that touching one section affects the whole. Individuals are but one intersection on the web: not separate, but crisscrossed by others and part of a larger, interrelated whole.

INFINITE REGRESS. Infinite regress is the third option in Agrippa's trilemma. It suggests that there is no way to ultimately secure our knowledge of the world. Rather, we can always ask questions and make challenges to evolving knowledge claims. Many philosophers, theologians, and scholars see infinite regress as a bad or undesirable option, but I have tried to argue for infinite regress in this book as an option that opens us onto our evolving, planetary contexts.

INSTRUMENTALIZATION. To treat something as an instrument is (in philosopher Immanuel Kant's terms) is to treat it as a thing or as means to an end. Instrumental thinking is at the heart of the modern Western scientific worldview that sees the rest of the natural world as mere dead matter for human projects. Under capitalism, many would argue, many human beings become instrumentalized and treated as mere labor or as resources, for example.

INTERSEXED. Being born intersexed is the state of being born neither as fully male nor female. There are various ways in which people are born intersexed, and as many as one in fifteen hundred human babies are born intersexed. Many in the American Medical Association and in the trans community are calling on doctors not to operate on intersexed children. These operations conform all life to the human construct of sexual dimorphism. Instead, intersex advocates call for allowing children to make choices about their own sex once they are able. For more information, visit the Intersex Society of North America's Web site: www.isna.org/.

INTERSTITIAL. This is a term often used in postcolonial thought that refers to the ways in which identities are formed through our interactions with others. Rather than understanding the self or any identity (or concept) as somehow starting with a solid, foundational, original essence and then coming in to interaction with others, this term suggests that all identities are formed, articulated, and made in the space in between the interactions of multiple others.

LGBT(Q). This is the acronym for lesbian, gay, bisexual, transgender, and queer or questioning. Some write it as GLBTQ and some add other letters in there, such as

I for intersexed or A for ally. However, in this text I just use LGBTQ with Q being a catchall category suggesting that all of our identities are somehow queer; there really is no normal (outside of statistical norms).

LINES OF FLIGHT. This is a phrase coined by Deleuze and Guattari to suggest imaginatively thinking about new ways of becoming into the future. Rather than getting caught in the same ideas, what we need are imaginative lines of flight that lead us toward different ways of relating to humans and earth others. Ideas, thoughts, and concepts should help us break out of habitual ways of being and becoming and think our lives anew.

LOGOCENTRISM. Logocentrism is literally the centrality of the word. This is what many post and deconstructive thinkers are trying to critique and do without. More broadly, it means that reality cannot be fixed by any foundational, original, or essential knowledge claims. It means that reality is constantly in flux and changing and our ideas about reality are never objective. When we take a concept, a knowledge claim, or a word to be central, the entire world is then organized in relation to that foundation and thus distorted.

MATERIALISM (MARXIST)/REDUCTIVE MATERIALISM. Materialism in the Marxist sense is the idea that some sort of material phenomenon, such as economics, can explain ideas, values, culture, and religious beliefs. Not all materialisms are reductive, but classic materialism in the sense Marx uses it is in direct opposition to Hegelian idealism. From this perspective, the ideal side of reality can be reduced to things like economic explanations or, in scientific materialism, everything can be reduced to genes, chemicals, or some other physical/material level.

MECHANISM. Mechanism is the predominant worldview of Western modern science. It is mostly associated with Isaac Newton's mechanical model of the universe and it suggests that mechanical forces can explain the whole cosmos. In other words, it is in direct opposition to something like the *élan vital*, which argues that the whole universe is alive in some sense. From the human perspective, this has contributed to the idea that the rest of the world is dead matter for human use.

MONISM. Monism is a philosophical claim about metaphysics that argues that there is no plurality of existence or dualism of material/matter and energy/spirit. It suggests that everything is of one kind. Though working against dualism, which monists see as destructive, this position has problems explaining diversity in ways that are not superficial: in other words, monism suggests that, though the world may appear to be diverse, at some level everything is the same.

MONOLOGICAL LOGIC. A monological logic imposes sameness over the entire face of the globe. In other words, it depends upon foundational thinking, which claims that "my" way is the true way and that all other ways must conform. Such a position toward human and earth other makes any kind of dialogical interaction impossible.

MULTINATURALISM. The concept of nature and what is natural has often been used as a leveling force. In other words, the idea is that while there are many different cultural expressions in the world and many different human perspectives, at the base of existence/reality all are of one nature. Multinaturalism suggests that there is no one concept of nature and that there are many natures all working toward realization in an unknown future.

MULTIPERSPECTIVALISM. This concept, also known as *syadvada*, suggests that reality is made up of multiple perspectives, none of which can grasp reality as it is. As such, the more perspectives one takes into account—human, animal, plant, and other—the better overall picture one might have of reality. This is a perspective that seeks to deny that all reality can be summed up under any one religion, concept, or perspective.

MULTISCALAR. Understanding that there are many levels of reality, none of which can be reduced to the other, and many different goals toward which these realities works calls for a multiscalar analysis of any given topic, ethical quandary, or situation. We can analyze things from the ecological, biological, genetic, chemical, religious, political, and psychological levels, and none of these can exhaust the reality of another level.

MULTIVERSE. The idea of a multiverse (rather than a universe) exists in many different religious traditions. In the West the idea is at least as old as Giordano Bruno, who argued that there may be many universes all existing at the same time. Some cosmologists posit a multiverse because it would seem that our universe is too unlikely to exist were there not many other attempts at such a universe. Furthermore, string theory, among others, suggests that there may be many universes running parallel to (and perhaps intersecting) our own. The idea of the multiverse is one way in which cosmologists have attempted to understand the competing laws of gravity and entropy.

NATURA NATURANS. This is the Latin term for "nature naturing." It is a term made famous by the seventeenth-century pantheist philosopher Spinoza. The idea is that, whatever it might be, nature is always in the process of becoming, thus the ever-creative-destructive process of nature on the move or an evental/processive understanding of reality.

NATURA NATURATA. This is the Latin term for "nature natured." Again made famous by Spinoza, it suggests that there is some aspect of nature (and the many different entities in nature) that remains constant through nature naturing. In this book I suggest that nature naturing is what best captures reality and that nature natured is something in Spinoza's thought that is left over from thinking about reality in terms of substances rather than events.

NEGATIVE THEOLOGIES. Negative theologies, similar to apophatic theologies and iconoclasm, suggest that we can only say what the divine is not. In other words, there are no positive statements that we can make about divinity, but we can say from our own experience that whatever God is must not be what we can imagine. This position relies more than apophatic and iconoclastic methodologies do on some sort of belief that there is an ultimate reality.

NETI-NETI. This is a Hindu concept meaning "not this, not that." It suggests that, whatever we may claim about the world, our claims should always be aware that we can never grasp the reality of any given entity, organism, or moment. Rather our words describe certain aspects of reality, but there is always a "not this, not that" quality to whatever we might say.

NONEQUILIBRIUM THERMODYNAMICS. The laws of thermodynamics were arrived at and applied to closed systems. In the reality in which we live, there are no closed systems, except perhaps the universe itself, but we don't know for sure. In other words, all entities live by exchanging energy and materials and thus live in a state of disequilibrium. Equilibrium means death or being absorbed by the world around you.

NONSUBSTANCE (NONSUBSTANTIVE) METAPHYSICS. Nonsubstance metaphysics do not understand essences, origins, or foundations as the basis of reality, but rather that change and process are somehow indicative of reality. Event-based and process-based metaphysics would be examples of nonsubstantive metaphysics.

ONTOLOGICAL. This word simply means "words about being." Much of philosophy has assumed that the understanding of what it means "to be" is what philosophy should attempt to articulate. However, the problem is that there are many different claims about being that lead some to suggest there is a gap between our knowledge and understanding of reality/being and what reality/being is in itself. Thus this book suggests that we can never get out of our contexts and knowledge to know something like true being. Instead, we must always talk about epistem-ontology or knowing and being together.

ORTHODOXY. Orthodoxy is most often used in relationship to the Orthodox Church, which is the Eastern interpretation of Christianity instead of the Latin/Western Catholic Church. Here, however, orthodoxy is used in a broader sense to connote the idea that there is a way of interpreting Scripture or tradition that is the true (orthodox) version. This book argues that there is no such thing as an orthodox interpretation of theologies or philosophies, but rather that there are and always have been many interpretations (polydoxy).

PERFORMATIVITY. This is a term from queer theory especially associated with Judith Butler. The idea is that there is nothing essential to identity, but every identity is a performance or an interpretation of various roles one has in society. It is not that we just create these performances out of nothing, but rather that histories and habits shape the way we act as male/female, human/animal, black/white, or teacher/student, for instance. In other words, we are performing roles all the time that are handed down to us, but there is nothing essential to our identities in any of these performances. Accordingly, it is in the way in which we perform on a daily basis that we have some freedom to choose how we will perform.

(PER)VERSION. A per/version simply means another version. Marcella Althaus-Reid suggests that all versions of identity, truth, and knowledge are just that: another version. There is no one version, but multiple versions for the way in which we live, understand, and act in the world.

PLACED-BASED ETHICS. A place-based ethic is an environmental ethic that suggests that the cause of our environmental and social ills is that modernity has led to some sort of place-less-ness in which no one lives in a given place and no one knows or has ethical responsibility for taking care of particular places. In this text I suggest that an ethic of place mimics foundational thinking and that we need an ethic of movement or a nomadic planetary ethic.

PLANETARITY. The planetary (or planetarity) is juxtaposed with globalization as a metaphor for how we might come together in differences in a way that does not overlook those differences. The concept, as it is used in this text, draws heavily from the postcolonial scholar and thinker Gayatri Spivak. Whereas globalization is the imposition of sameness over the face of the planet, planetarity is a way to think about how we are codefined and come together because of, with, and through our differences.

PLURISINGULAR. This concept draws heavily from the works of French philosopher Jean-Luc Nancy and theologian Catherine Keller. It suggests that reality/identity is always already multiple and interconnected. It seeks to avoid the sameness problem of holism, while also avoiding individualism or atomism, which suggests that entities are somehow disconnected from one another.

POLYAMORY (POLYAMORY OF PLACE). This is the idea and practice of multiple loves. In sharp contrast to monogamy, which may be historically and politically tied up with capitalism, heterosexism, and patriarchy, polyamory suggests that it is possible to have multiple, open, loving (and sexual) relationships. A polyamory of place is used in this book to critique place-based thinking in environmental ethics and suggests that we must learn to love multiple places all at once.

POLYDOXY. The idea of polydoxy is in direct opposition to any notion of an orthodox interpretation of a given tradition. Polydoxy suggests that there has never been a single version of reality, but every history and tradition has always already involved multiple interpretations and voices. Most recently, this discourse has developed in opposition to the movement known as radical orthodoxy.

POSTCOLONIAL. Postcolonial studies and scholarship focus on the ways in which colonization continues despite the fact that physical occupation of colonies by empires is, for the most part, over. These studies focus on how identities and traditions are always already codefined; thus there are no pure traditions. Postcolonial scholarship is particularly critical of any idea of a single world history and seeks to uncover the hidden or lost voices of the past, using feminist, race, class, sexual orientation, and other positionalities as points of departure for alternative ways to analyze history and politics.

POSTFOUNDATIONAL. Postfoundationalism is an epistemological stance that thinks about what knowledge might mean if we do not use foundational understandings (as in Agrippa's trilemma). It is also distinct from antifoundationalism, which suggests that there is no way to discern between different knowledge claims or assumes complete relativity.

POSTMODERN(ISM). In contrast to modernity, which suggests that there is a single history, reality, and movement toward progress in the world, and that human reason can know the world as it is, postmodernity suggests that there is no single story, but rather there are multiple histories, multiple realities, and our knowledge is perspectival, never total, and never knowledge of the world as it is in itself. Poststructuralism, deconstructionism, postcolonialism, feminist studies, queer studies, and critical race theory are all some examples of postmodern thought.

PRAGMATISTS (AMERICAN). Pragmatism is an early twentieth-century American philosophical movement that operates under several assumptions that are in opposition to dualistic thought and modern thought. It moves questions about knowledge (including religious and scientific knowledge) away from questions about ontology and metaphysics (ultimate reality) and toward questions about ethics. The question then becomes how various ways of knowing "take place" in the world and affect the lives and bodies of human and earth others.

PRATĪTYASAMUTPĀDA OR PATICCASAMUPPĀDA (DEPENDENT COARISING). This is the idea that everything is connected. One cannot separate things out one from another, nor one's self from human and earth others (present, past, and future). This is one of the major tenets of Buddhism that is quite different from the individualistic identities and substance-based metaphysics of, say, the Abrahamic faiths. In this perspective there is no self or other that is separate from other selves and others; thus change, interaction, and interrelatedness are what mark the true nature of reality.

PRECAUTIONARY PRINCIPLE. This is an ethical stance, particularly in environmental ethics, that suggests we ought not adopt a new technology or act if we do not have enough information to know how that technology or action will fully affect ecosystems, future generations, and other animals. The problem with this position is that there is never full information and we always act from positions of uncertainty.

PROCESS THOUGHT (METAPHYSICS). Process thought is a nonsubstance, process-based metaphysics developed by the philosopher and mathematician Alfred North Whitehead in the twentieth century. He developed this philosophy through his interpretation of quantum physics. In general, the philosophy suggests that movement, interconnection, and change mark reality more than any sort of substance or essence-based metaphysic. This has been an important source for thinking about God as immanent and for a religiously based environmental ethic.

PTOLEMAIC UNIVERSE (ARISTOTLE'S UNIVERSE). This is the view of the universe developed by Ptolemy (and held by Aristotle before that) that is also known as the geocentric or earth-centered understanding of the universe. The earth is at the center with all of the planets and the sun revolving around the earth; beyond this is the realm of the fixed stars and beyond that the *aether* (where the Unmoved Mover that sets the whole cosmos into motion resides). It was this worldview that Christian thought (among others) was developed through for over one thousand years, challenged by Copernicus and Galileo.

REGIMES OF TRUTH. This is an idea developed by French philosopher Michel Foucault that suggests that whatever truth is, it acts like habits that guide us into certain ways of being in the world. In other words, the mechanical understanding of the universe developed by Newton helps to create the world in the form of the technological and industrial revolution. This suggests that truth is more about ethics and action than metaphysics or ontology.

REIFICATION. Reification is what happens when we assume that our knowledge, language, and concepts can capture reality fully. Such capture turns reality into things for human use and helps to make the rest of the world an instrument toward human ends.

RELATIVISM. Relativism is the idea that knowledge is completely perspectival and based upon one's culture and life experience. From a strictly relativist perspective, no two competing ideas can really be debated or compared because one idea is just as good as any other. This text argues for contextuality, which is a position between universalism and relativity.

RENAISSANCE. The Renaissance, or rebirth, is usually associated with the historical period in Western history in which ancient Greek and Roman culture was rediscovered by the Latin West (beginning in the twelfth century). The idea of the renaissance, following the Dark Ages, helps to promote the idea that the West is the heir of the Greco-Roman world and also ignores the contribution of Islamic scholars to natural philosophy and science during the Golden Age of Islam.

RHIZOMES. Rhizomes or a rhizomatic ontology/epistemology is a metaphor proposed by Gilles Deleuze and Felix Guattari to replace arboreal thinking. Arboreal thought is the type of thinking that looks for origins, foundations, and roots in reality and implies that there is a definite, discernible growth toward a specific direction. Rhizomatic thought, on the other hand, has no origin or goal, but rather can shoot off in many different directions.

SOLIPSISM. Solipsism is a process whereby an argument is self-referential and closed in on itself. The idea is that one's self is the only real reference point for reality, and no other reference point can be trusted. I use identity solipsism in this book to refer to the idea that one's identity is an essential substance formed before interacting with human and earth others.

STRATEGIC ESSENTIALISMS. A strategic essentialism is a tactic/method in postcolonial theory for deciding on ethical and political action in a world that is postmodern and contextual. First used by Gayatri Spivak, the term seeks to address the critique that postmodern and post thinking in general leave no way to determine what is good/bad, right/wrong, true/false, and thus makes ethics/politics impossible. This is one method for doing ethics from a postfoundational perspective.

SUBSTANCE-BASED METAPHYSICS (SUBSTANTIAL/SUBSTANTIVE METAPHYSICS). A substance-based metaphysic suggests that there are essences, ultimate goals, or ultimate forms of reality toward which reality moves and progresses. It can also be used in the sense of describing the Newtonian mechanical universe where the world is ultimately made up of billiard ball like atoms: if we can just understand and control these billiard balls, then we can control nature. Substance-based thinking is challenged by process and event-based ontologies and metaphysics that suggest change is the true nature of reality.

SUI GENERIS. This is a Latin term meaning "self-sufficient" or "of its own kind." In philosophical and epistemological terms, an idea or concept that is sui generis is foundational, cannot be reduced, and needs no explanation. In other words, it is one of the foundational ways in which knowledge systems anchor themselves in reality. It is one way of responding to Agrippa's trilemma.

SYADVADA (RELATIVE PERSPECTIVES). This is a Jain concept meaning that there are many and multiple perspectives of the world, none of which capture all of reality (cf. *Anekanta*), but all of which together make up the whole of reality. Reality, according to this doctrine, is multiperspectival through and through.

TECHNOCAPITALISM. Technocapitalism is used by some to describe the current, dominant theory of Western culture. In other words, the idea is that technology and economics have all the answers to our problems. Whereas many people look to religion and other modes of thought for answers to the problems presented by life, those in the globalized world increasingly look to technology and science for answers, solutions, and even salvation. The capitalism part simply indicates that technological solutions are currently directed by the concerns of capitalism: that is, whether or not a solution is good often depends on economic rather than religious, social, environmental, or other measures.

TECHNOLOGIES OF MEANING. This phrase is an attempt in this book to expand our understanding of technology to include language, ideas, values, and religious beliefs and meaning-making systems. In other words, our meaning-making practices shape our bodies and other bodies in the world around us. Further, we are shaped by histories of technologies of meaning. Suggesting that meaning is a form of technology means that we ought to analyze the effects of meaning and retool meanings that we determine to have adverse effects on the planetary community.

TELEOLOGY. This is the idea of end-oriented solutions to knowledge and meaning. Aristotle is often considered to be the original teleological thinker. Teleology provides a method for knowledge and ethics that asks the question: toward what end are things for? Many forms of postmodern and deconstructive thought suggest that teleology has taken us away from ethical concerns and that it leads us toward a future idea of how we think things ought to be that ignores the injustices, complexities, and messiness of the present moment.

TERRITORIALIZATION (RETERRITORIALIZATION/DETERRITORIALIZATION). These are terms made popular by the French philosophers Gilles Deleuze and Felix Guattari. They suggest that our role as thinking creatures is to think where we are going and what we are doing anew. We must always destabilize concepts so that we recognize that they never capture reality, yet as meaning-making creatures we always have to territorialize in order to live life. This idea is meant to help us continue to never stop thinking anew and to realize that whatever we think reality, identity, or value might be, there is always more.

THIRD/FOURTH GENDERS (HIJRA, WARIA, NADLE, TWO SPIRITS). Third, fourth, and two-spirit genders exist in many cultures and traditions. These identities trip up the concept of gender dimorphism that is imposed on most of the world through Western modes of thinking. Most cultures and traditions have more than two genders/sexes, and third genders, two spirits, or fourth genders are ways of describing the many different ways we can be in the world. In other words, it is not clear that the world can be divided into male/female.

TRADITIONAL ECOLOGICAL KNOWLEDGE (TEK). Traditional ecological knowledge or TEK is a term that is used to describe the different options of knowing material reality from that of modern, Western science. In fact, modern Western science draws off of traditional ecological knowledges and uses such knowledge. TEK is especially relevant in arguing against pharmaceutical companies that often rely on local people's knowledge of plants to patent certain chemical properties and then make them the property of the pharmaceutical industry. In other words, TEK is the way in which many Western pharmaceutical companies learn about the benefits of plants for various maladies, yet claim such knowledge as their own.

TRANSCENDENCE. Transcendence is the option that most meaning-making practices take to articulate that whatever reality might be is beyond what we know and sense of the world. In other words, whereas the reality of the world might be one of change, transition, suffering, and death, there is some sort of parallel transcendent reality that is constant and universal. This notion of transcendence has been used to describe ultimate reality, the divine plan, and even future states of being that are more ecologically sound and socially just. In this text I argue (along with Marx, Nietzsche, Foucault, Derrida and postcolonial and postmodern thought) that whatever we might think of as divinity, utopia, or alternate realities ought to be thought of in terms of immanence rather than transcendence.

TRICKSTER. The trickster is a mythological and narrative figure that appears in many different religious traditions, but most notably in indigenous traditions. This figure, often conceived as an animal with human qualities (such as a raven or coyote), is that which navigates between the boundaries of human/nonhuman, living/dead, machine/organism (cyborg), and many other dualistic categories into which we tend to separate our realities.

UNIVERSALISM/ UNIVERSALITY. Universal thinking is a form of foundational thought (essential, substantive, original, teleological) that suggests that human reason or experience can know reality as it is for all times and all places. In this book I suggest that universal thought breaks apart under the pressure of a 13.7-billion-year universe and a 4.5-billion-year process of planetary evolution in which many new ideas, natural laws, meanings, realities, organisms, entities, and values have emerged.

QUANTUM REALITY/PHYSICS. Quantum reality and physics should be separated from the scientists that study the quantum level. Many of these scientists still adhere to a mechanistic and reductive scientific method and suggest that the quantum reality is the base level by which all life can be described. There are many insights in quantum science that lead to the dissolution of substantive metaphysics, such as the inseparability of matter-energy and space-time, the refutation that there is any base of reality (quantum physics has identified subatomic particles such as neutrinos, quarks, muons, and neutrinos), and suggests that there will never be any bottom to reality but more and more energy events. Because of this, some philosophers of science have called quantum physics a postmodern science. It upsets, from a scientific perspective, the same ideas of objectivity, substance, and essences that postmodern thought upsets. To be sure, the science of quantum physics is still based very much in mathematics (which can be a form of foundational thought), but the challenges raised by quantum physics about our everyday experience of reality can be described as postmodern challenges. In this text I suggest that nonsubstantive modes of understanding the world (such as those found in Vedic cultures) have influenced the ways in which scientists look at the world. This is not to suggest that we create the world through ideas (a form of Hegelian idealism), but that our metaphors of reality change the ways we perceive and look at reality. If the reader is interested in this, he may want to monitor the changes in our scientific understanding of the world that are emerging from the Hadron Collider: www.lhc.ac.uk/. Furthermore, it is worth noting that the quantum reality, which the Hadron Collider is examining, was influenced by such thinkers as Einstein and Bohr, who were very much influenced by Eastern (nonsubstantive) ways of looking at the world.

QUEER THEORY/QUEERING. Queer theory is the critical philosophical thought that was begun with Michel Foucault and has been most taken up by contemporary thinkers such as Judith Butler. This theory assumes that the assumptions of heterosexuality, monogamy, gender and sexual dimorphism, and other norms are not in any way natural but created through time, tradition, politics, and power dynamics. As such, queer theory challenges all ideas that purport to be natural, universal, and given. In a simple sense, queer theory suggests that reality is much stranger than any thought, idea, system, or belief can capture.

WORKS CITED

Abram, David. *Becoming Animal: An Earthly Cosmology*. New York: Vintage, 2010.
Alcoff, Linda, and Elizabeth Potters, eds. *Feminist Epistemologies*. New York: Routledge, 1993.
Althaus-Reid, Marcella. *Indecent Theology: Theological Perversions in Sex, Gender, and Politics*. New York: Routledge, 2000.
Anderlini-D'Onofrio. *Gaia and the New Politics of Love: Notes for a Poly Planet*. Berkeley: North Atlantic, 2009.
Anker, Michael. *The Ethics of Uncertainty: Aporetic Openings*. New York: Atropos, 2009.
Anzaldúa, Gloria. *Borderlands: La Frontera. The New Mestiza*. San Francisco: Aunt Lute, 1987.
Arendt, Hannah. *Eichmann in Jerusalem: A Report on the Banality of Evil*. New York: Viking, 1963.
Armour, Ellen, and Susan St. Ville, eds. *Bodily Citations: Religion and Judith Butler*. New York: Columbia University Press, 2006.
Asad, Talal. *Formations of the Secular: Christianity, Islam, Modernity*. Stanford: Stanford University Press, 2003.
Barad, Karen. *Meeting the Universe Halfway: Quantum Physics and the Entanglement of Matter and Meaning*. Durham: Duke University Press, 2007.
Barbour, Ian G. *Religion and Science: Historical and Contemporary Issues*. San Francisco: HarperCollins, 1990.

Bauman, Whitney. "The Eco-Ontology of Social/ist Eco-Feminist Thought." *Environmental Ethics* 29 (Fall 2007): 279–298.

———. "Fashioning a Persuasive Environmental Ethic: Thinking Without Surface and Depth." *Ecozona* 2, no. 2 (2011): 17–39.

———. "Religion, Science and Nature: Shifts in Meaning on a Changing Planet." *Zygon: Journal of Religion and Science* 46, no. 4 (2011): 777–792.

———. "Technology and the Polytheistic Mind: From the Truth of the Global to Planetary 'Lines of Flight.'" *Dialog: A Journal of Theology* 50, no. 4 (2011): 344–353.

———. *Theology, Creation and Environmental Ethics: From Creatio ex Nihilo to Terra Nullius*. New York: Routledge, 2009.

Bauman, Zygmunt. *Globalization: The Human Consequences*. New York: Columbia University Press, 1998.

———. *Liquid Modernity*. Cambridge: Polity, 2000.

Bedau, Mark, and Paul Humphreys. *Emergence: Contemporary Readings in Philosophy and Science*. Boston: MIT Press, 2008.

Bellah, Robert. *Religion in Human Evolution: From the Paleolithic to the Axial Age*. Cambridge: Harvard University Press, 2011.

Bennett, Jane. *Vibrant Matter: A Political Ecology of Things*. Durham: Duke University Press, 2010.

Bergmann, Sigurd, and Tore Sager, eds. *The Ethics of Mobilities: Rethinking Place, Exclusion, Freedom and the Environment*. Aldershot: Ashgate, 2008.

Bergson, Henri. *Creative Evolution*. New York: Modern Library, 1944.

———. *The Two Sources of Morality and Religion*. London: MacMillan, 1935.

Berry, Wendell. "Feminism, the Body, and the Machine." *Cross Currents* 53, no. 1 (Spring 2003): 32–48.

Bhabha, Homi. *The Location of Culture*. London: Routledge, 1994.

Boddice, Rob, ed. *Anthropocentrism: Humans, Animals, Environments*. Leiden: Brill, 2011.

Boellstorff, Tom. *The Gay Archipelago: Sexuality and the Nation in Indonesia*. Princeton: Princeton University Press, 2005.

———."Playing Back the Nation: Waria, Indonesian Transvestites." *Cultural Anthropology* 19, no. 2 (May 2004): 159–195.

Boesel, Chris, and Catherine Keller, eds. *Apophatic Bodies: Negative Theology, Incarnation, and Relationality*. New York: Fordham University Press, 2010.

Bourdieu, Pierre. *The Logic of Practice*. Stanford: Stanford University Press, 1980.

Braidotti, Rosi. *Transpositions*. Cambridge: Polity, 2006.

Brennan, Teresa. *Exhausting Modernity: Grounds for a New Economy*. New York: Routledge, 2000.

———. *Globalization and Its Terrors*. New York: Routledge, 2003.

Brooke, John Hedley. *Science and Religion: Historical Perspectives*. Cambridge: Cambridge University Press, 1991.
Butler, Judith. *Bodies That Matter: On the Discursive Limits of Sex*. New York: Routledge, 1993.
———. *The Psychic Life of Power: Theories in Subjection*. Stanford: Stanford University Press, 1997.
Calhoun, Craig, Mark Juergensmeyer, and Jonothan van Antwerpen. *Rethinking Secularism*. New York: Oxford University Press, 2011.
Campbell, Mary Baine. *Wonder and Science: Imagining Worlds in Early Modern Europe*. Ithaca: Cornell University Press, 1999.
Clarke, J. J. *Oriental Enlightenment: The Encounter Between Asian and Western Thought*. New York: Routledge, 1997.
Clayton, Philip, and Paul Davies. *The Re-Emergence of Emergence: The Emergentist Hypothesis from Science to Religion*. Oxford: Oxford University Press, 2006.
Code, Lorraine. *Ecological Thinking: The Politics of Epistemic Location*. Oxford: Oxford University Press, 2006.
Colebrook, Claire. *Gilles Deleuze*. New York: Routledge, 2001.
Connolly, William. *A World of Becoming*. Durham: Duke University Press, 2011.
Cooper, David E., and Joy A. Palmer, eds. *Spirit of the Environment*. New York: Routledge, 1998.
Cruz-Malave, Arnaldo, and Martin F. Manalansan, eds. *Queer Globalizations: Citizenship and the Afterlife of Colonialism*. New York: New York University Press, 2002.
Deacon, Terrence. *Incomplete Nature: How Mind Emerged from Matter*. New York: Norton, 2011.
———. *The Symbolic Species: The Coevolution of Language and the Brain*. New York: Norton, 1997.
De Chardin, Teilhard. *The Phenomenon of Man*. New York: Harper and Row, 1959.
De La Torre, Miguel, and Albert Hernandez. *The Quest for the Historical Satan*. Minneapolis: Fortress, 2011.
Deleuze, Gilles, ed. *Expressionism in Philosophy: Spinoza*. Brooklyn: Zone, 1992.
———. *The Fold: Leibniz and the Baroque*. Minneapolis: University of Minnesota Press, 1992.
Deleuze, Gilles, and Felix Guattari. *What Is Philosophy?* New York: Columbia University Press, 1994.
———. *A Thousand Plateaus: Capitalism and Schizophrenia*. Minneapolis: University of Minnesota Press, 1987.
Derrida, Jacques. *Of Grammatology*. Trans. Gayatri Spivak. Baltimore: John Hopkins University Press, 1974.

———. *Specters of Marx: The State of the Debt, The Work of Mourning and the New International*. New York: Routledge, 1994.

———. *Writing and Difference*. London: Routledge, 1967.

Derrida, Jacques, and Gianni Vattimo. *Religion: Cultural Memory in the Present*. Stanford: Stanford University Press, 1996.

Dieter, Richard C. *The Death Penalty in Black and White: Who Lives, Who Dies, and Who Decides*. Washington, DC: Death Penalty Information Center, 1998.

Dupre, Louis. *Passage to Modernity: An Essay in the Hermeneutics of Nature and Culture*. New Haven: Yale University Press, 1993.

Durkheim, Emile, ed. *The Elementary Forms of Religious Life*. New York: Free Press, 1995.

Easton, Dossie, and Janet Hardy. *The Ethical Slut: A Practical Guide to Polyamory, Open Relationships, and Other Adventures*. Berkeley: Ten Speed, 1997.

Eckstein, Walter. "The Religious Elements in Spinoza's Thought." *Journal of Religion* 23, no. 3 (July1943): 153–163.

Engels, Friedrich, ed. *The Origin of the Family, Private Property and the State*. New York: Penguin, 1972.

Feuerbach, Ludwig. *Lectures on the Essence of Religion*. Trans. Ralph Manheim. New York: Harper and Row, 1967.

Foucault, Michel. *The History of Sexuality*, vol. 1: *An Introduction*. New York: Random House, 1978.

———. *Power/Knowledge: Selected Interviews and Other Writings, 1972–1977*. New York: Random House, 1972.

Gaard, Greta. "Toward a Queer Ecofeminism." *Hypatia* 12, no. 1 (February 1997): 114–137.

Gabilondo, Jose. "Asking the Straight Question: How to Come to Speech in Spite of Conceptual Liquidation as a Homosexual." *Wisconsin Women's Law Journal* 21, no. 2 (Summer 2006): 101–146.

Gatens, Moria. "Feminism as 'Password': Rethinking the 'Possible' with Spinoza and Deleuze." *Hypatia* 15, no. 2 (Spring 2000): 59–75.

Gelobter, Michael, Michael Dorsey, and Leslie Fields, Tom Goldtooth, Anuja Mendiratta, Richard Moore, Rachel Morello-Frosch, Peggy Shepard, and Gerald Torres. *The Soul of Environmentalism*. Oakland: Redefining Progress, 2005.

Goldstein, E. Bruce. *Introduction to Perception*. 8th ed. Belmont, CA: Wadsworth, 2010.

Goodchild, Philip. *Theology of Money*. Durham: Duke University Press, 2009.

Goodenough, Ursula, and Terrence. Deacon. "The Sacred Emergence of Nature." In Philip Clayton, ed., *The Oxford Handbook of Religion and Science*, 853–871. Oxford: Oxford University Press, 2006.

Grant, Ian Hamilton. *Philosophies of Nature After Schelling*. New York: Continuum, 2008.

Grau, Marion. *Of Divine Economy: Refinancing Redemption*. New York: T&T Clark, 2006.
Green, Garrett. *Theology, Hermeneutics, and Imagination: The Crisis of Interpretation at the End of Modernity*. Cambridge: Cambridge University Press, 2000.
Griffin, David Ray. *The Reenchantment of Science*. Albany: State University of New York Press, 1998.
Gudorf, Christine. "The Erosion of Sexual Dimorphism: Challenges to Religion and Religious Ethics. *Journal of the American Academy of Religion* 69, no. 4 (2001): 863–891.
Haag, James, and Whitney Bauman. "De/Constructing Transcendence: The Emergence of Religion Bodies." In David Cave and Rebecca Norris, eds., *The Body and Religion: Modern Science and the Construction of Religious Meaning*, 37–55. Leiden, Brill, 2012.
Haraway, Donna. *The Companion Species Manifesto: Dogs, People and Significant Otherness*. Chicago: Prickly Paradigm, 2003.
———. *Simians, Cyborgs, and Women: The Reinvention of Nature*. New York: Routledge, 1991.
———. "Situated Knowledges: The Science Question in Feminism and the Privilege of Partial Perspective." *Feminist Studies* 14, no. 3 (1988): 575–599.
———. *When Species Meet (PostHumanities)*. Minneapolis: University of Minnesota Press, 2007.
Harding, Sandra. *Is Science Multicultural? Postcolonialisms, Feminisms, and Epistemologies*. Bloomington: Indiana University Press, 1998.
———. *The Postcolonial Science and Technology Studies Reader*. Durham: Duke University Press, 2011.
Hargens, Sean, and Michael Zimmerman. *Integral Ecology: Uniting Multiple Perspectives on the Natural World*. Boston: Integral, 2009.
Harvey, Van. *Feuerbach and the Interpretation of Religion*. Cambridge: Cambridge University Press, 1995.
Hefner, Philip. *Technology and Human Becoming*. Minneapolis: Fortress, 2003.
Heidegger, Martin. *The Question Concerning Technology and Other Essays*. New York: Harper and Row, 1977.
Heise, Ursula. *Sense of Place and Sense of Planet: The Environmental Imagination of the Global*. New York: Oxford University Press, 2008.
Herzogenrath, Bernd, ed. *Deleuze/Guattari and Ecology*. New York: Palgrave Macmillan, 2009.
Heyd, Thomas, ed. *Recognizing the Autonomy of Nature*. New York: Columbia University Press, 2005.
Hoad, Neville. "Run Caster Semenya, Run: Nativism and the Translations of Gender Variance." *Safundi: The Journal of South African and American Studies* 11, no. 4 (October 2010): 397–405.

Holmes, Barbara. *Race and the Cosmos: An Invitation to View the World Differently*. Harrisburg, PA: Trinity, 2002.

Horkheimer, Max, and Theodor Adorno, eds. *Dialectic of the Enlightenment*. Trans. Edmund Jephcott. Stanford: Stanford University Press, 2007.

Hyde, Lewis. *Trickster Makes This World: Mischief, Myth and Art*. New York: North Point, 1998.

Jackson, William. *The Wisdom of Generosity: A Reader in American Philanthropy*. Waco: Baylor University Press, 2008.

Kaiser, David. *How the Hippies Saved Physics: Science, Counterculture, and the Quantum Revival*. New York: Norton, 2011.

Kaufman, Gordon. *An Essay on Theological Method*. Atlanta: Scholars Press, 1975.

———. *In Face of Mystery: A Constructive Theology*. Cambridge: Harvard University Press, 1993.

———. *In the Beginning . . . Creativity*. Minneapolis: Fortress, 2004.

———. *The Theological Imagination: Constructing the Concept of God*. Louisville: Westminster John Knox, 1981.

Kearney, Richard. *Anatheism: Returning to God After God*. New York: Columbia University Press, 2010.

Keller, Catherine. *Apocalypse Now and Then: A Feminist Guide to the End of the World*. Boston: Beacon, 1996.

———. *Face of the Deep: A Theology of Becoming*. New York: Routledge, 2003.

———. *God and Power: Counter-Apocalyptic Journeys*. Minneapolis: Fortress, 2005.

Keller, Catherine, and Anne Daniel. *Process and Difference: Between Cosmological and Poststructuralist Postmodernisms*. Albany: State University of New York Press, 2002.

Keller, Catherine, and Laurel Kearns. *EcoSpirit: Religions and Philosophies for the Earth*. New York: Fordham University Press, 2007.

Keller, Catherine, Michael Nausner, and Mayra Rivera, eds. *Postcolonial Theologies: Divinity and Empire*. St. Louis: Chalice, 2004.

Keller, Catherine, and Laurel C. Schneider, eds. *Polydoxy: A Theology of Multiplicity and Relation*. New York: Routledge, 2011.

Kuhn, Thomas. *The Structure of Scientific Revolutions*. Chicago: University of Chicago Press, 1962.

Kung, Hans. *Global Responsibility: In Search of a New World Ethic*. New York: Continuum, 1993.

Latour, Bruno. *Politics of Nature: How to Bring Sciences Into Democracy*. Cambridge: Harvard University Press, 2004.

———. *Reassembling the Social: An Introduction to Actor-Network Theory*. Oxford: Oxford University Press, 2005.

———. *We Have Never Been Modern*. Cambridge: Harvard University Press, 1993.

Lewellyn, John. *Margins of Religion: Between Kierkegaard and Derrida*. Bloomington: Indiana University Press, 2008.

Lorde, Audre. *Sister Outsider: Essays and Speeches*. Berkeley: Crossing, 1984.

Lovejoy, C. Owen. "Reexamining Human Origins in Light of Ardipithecus Ramidus." *Science* 326, no. 74 (October 2009): 74–74e8.

Lovelock, James. *Gaia: A New Look at Life on Earth*. Oxford: Oxford University Press, 1979.

Marable, Manning, Ian Steinberg, and Keesha Middlemass, eds. *Racializing Justice, Disenfranchising Lives: The Racism, Criminal Justice, and Law Reader*. New York: Palgrave, 2007.

McIntosh, Peggy. "White Privilege and Male Privilege: A Personal Account of Coming to See Correspondences Through Work in Women's Studies." Working Paper no. 189. Wellesley: Center for Research on Women at Wellesley College, 1988.

McKibben, Bill. *The End of Nature*. New York: Random House, 1989.

Merchant. Carolyn. *The Death of Nature: Women, Ecology and the Scientific Revolution*. New York: HarperCollins, 1980.

———. *Reinventing Eden: The Fate of Nature in Western Culture*. New York: Routledge, 2003.

Mignolo, Walter. *The Darker Side of the Renaissance: Literacy, Territoriality, and Colonization*. Ann Arbor: University of Michigan Press, 2003.

Moore, Stephen D., and Mayra Rivera, eds. *Planetary Loves: Spivak, Postcoloniality, and Theology*. New York: Fordham University Press, 2011.

Mortimer-Sandilands, Catriona, and Bruce Erickson, eds. *Queer Ecologies: Sex, Nature, Politics, Desire*. Bloomington: Indiana University Press, 2010.

Morton, Timothy. *The Ecological Thought*. Cambridge: Harvard University Press, 2010.

———. *Ecology Without Nature: Rethinking Environmental Aesthetics*. Cambridge: Harvard University Press, 2009.

Munro, Brenna. "Caster Semenya: God's and Monsters." *Safundi: The Journal of South African and American Studies* 11, no. 4 (2010): 383–396.

Nancy, Jean-Luc. *Being Singular Plural*. Stanford: Stanford University Press, 2000.

———. *The Creation of the World or Globalization*. Albany: State University of New York Press, 2007.

Nelson, Cary, and Lawrence Gross, eds. *Marxism and the Interpretation of Culture*. London: Macmillan, 1988.

Nelson, Robert. *The New Holy Wars: Economic Religion vs. Environmental Religion in Contemporary America*. University Park: Penn State University Press, 2009.

Nietzsche, Friedrich, ed. *Beyond Good and Evil: Prelude to a Philosophy of the Future*. Trans. Walter Kaufmann. New York: Vintage, 1966.

Nixon, Rob. *Slow Violence and the Environmentalism of the Poor*. Cambridge: Harvard University Press, 2011.

Noble, David. *The Religion of Technology: The Divinity of Man and the Spirit of Invention*. New York: Penguin, 1999.

Numbers, Ronald L. *Galileo Goes to Jail and Other Myths About Science and Religion*. Cambridge: Harvard University Press, 2010.

O'Brien, Kevin. *An Ethics of Biodiversity: Christianity, Ecology, and Variety of Life*. Washington, DC: Georgetown University Press, 2010.

O'Reilly, James, Sean O'Reilly, and Richard Sterling, eds. *The Ultimate Journey: Inspiring Stories of Living and Dying*. San Francisco: Travelers Tales, 2000.

Peterson, Anna. *Being Human: Ethics, Environment, and Our Place in the World*. Berkeley: University of California Press, 2001.

Pinker, Steven. *The Better Angels of Our Nature: Why Violence Has Declined*. New York: Viking, 2011.

Plumwood, Val. *Environmental Culture: The Ecological Crisis of Reason*. New York: Routledge, 2001.

———. "Shadow Places and the Politics of Dwelling." *Australian Humanities Review* 44 (2008): 139–150.

Proctor, James, ed. *Science, Religion and the Human Experience*. New York: Oxford University Press, 2005.

Rorty, Richard. *Philosophy and Social Hope*. London: Penguin, 1999.

Roughgarden, Joan. *Evolution's Rainbow: Diversity, Gender, and Sexuality in Nature and People*. Berkeley: University of California Press, 2004.

Rubenstein, Mary Jane. *Strange Wonder: The Closure of Metaphysics and the Opening of Awe*. New York: Columbia University Press, 2008.

Rue, Loyal. *Nature Is Enough: Religious Naturalism and the Meaning of Life*. Albany: State University of New York Press, 2011.

Ruether, Rosemary Radford. *Christianity and the Making of the Modern Family: Ruling Ideologies, Diverse Realities*. Boston: Beacon, 2000.

———. *Gaia and God: An Ecofeminist Theology of Earth Healing*. New York: HarperCollins, 1992.

Ruether, Rosemary Radford, and Marion Grau. *Interpreting the Postmodern: Responses to Radical Orthodoxy*. New York: Continuum, 2006.

Sabatino, Charles. "Projection as Symbol: Rethinking Feuerbach's Criticism. *Encounter* 48, no. 2 (Spring 1987): 179–193.

Said, Edward. *Orientalism*. New York: Vintage, 1979.

Scheibinger, Londa. *Nature's Body: Gender in the Making of Modern Science*. Trenton, NJ: Rutgers University Press, 2004.

Schneider, Eric, and Dorion Sagan. *Into the Cool: Energy Flow, Thermodynamics, and Life.* Chicago: University of Chicago Press, 2005.

Schneider, Laurel. *Beyond Monotheism: A Theology of Multiplicity.* New York: Routledge, 2008.

Sedgwick, Eve Kosofsky. *Epistemology of the Closet.* Berkeley: University of California Press, 1991.

Shellenberger, Michael, and Ted Nordhaus. "The Death of Environmentalism." Environmental Grantmaker's Association Position Paper, October 2004.

Singer, Dorthea. "The Cosmology of Giordano Bruno (1548–1600)." *Isis* 32, no. 2 (June 1941): 187–196.

———. *Giordano Bruno: His Life and Thought.* New York: Henry Schuman, 1950.

Soper, Kate, Martin Ryle, and Lyn Thomas, eds. *The Politics and Pleasure of Consuming Differently.* New York: Palgrave, 2009.

Spade, Dean. *Normal Life: Administrative Violence, Critical Trans Politics, and the Limits of Law.* Cambridge: Southend, 2011.

Spencer, Daniel. *Gay and Gaia: Ethics, Ecology, and the Erotic.* Cleveland: Pilgrim, 1996.

Spinoza, Benedict de, ed. *Ethics.* London: Penguin, 1996.

———. *Parallel Passages Between Giordano Bruno and Benedict de Spinoza.* Whitefish, MO: Kessinger, 2006.

Spivak, Gayatri. *Death of a Discipline.* New York: Columbia University Press, 2003.

Stryker, Susan. *Transgender History.* Berkeley: Seal, 2008.

Swimme, Brian, and Thomas Berry. *The Universe Story: From the Primordial Flaring Forth to the Ecozoic Era.* San Francisco: Harper, 1994.

Swimme, Brian, and Mary Evelyn Tucker. *Journey of the Universe.* New Haven: Yale University Press, 2011.

Taylor, Bron. *Dark Green Religion: Nature Spirituality and the Planetary Future.* Berkeley: University of California Press, 2009.

Taylor, Charles. *A Secular Age.* Cambridge: Harvard University Press, 2007.

Taylor, Mark. *Erring: A Postmodern A/theology.* Chicago: University of Chicago Press, 1984.

Tucker, Mary Evelyn. *Worldly Wonder: Religions Enter Their Ecological Phase.* Chicago: Open Court, 2003.

Tucker, Mary Evelyn, and John Grim, eds. *Religions of the World and Ecology.* 9 vols. Boston: Harvard University Press, 1997–2004.

Turing, Alan. "Computing Machinery and Intelligence." *Mind* 59, no. 236 (October 1950): 433–460.

Tweed, Thomas. *Crossing and Dwelling: A Theory of Religion.* Cambridge: Harvard University Press, 2006.

Van Huyssteen, Wentzel J. *The Shaping of Rationality: Toward Interdisciplinarity in Theology and Science*. Grand Rapids: Eerdmans, 1999.

Weber, Max, ed. *The Protestant Ethic and the Spirit of Capitalism*. Oxford: Oxford University Press, 2008.

Weststeijn, Thijs. "Spinoza sinicus: An Asian Paragraph in the History of the Radical Enlightenment." *Journal of the History of Ideas* 68, no. 4 (2007): 538.

Whitehead, Alfred North. *Process and Reality*. New York: Free Press, 1978.

Žižek, Slavoj. *The Fragile Absolute; or, Why Is the Christian Legacy Worth Fighting For?* London: Verso, 2000.

———. *In Defense of Lost Causes*. London: Verso, 2008.

INDEX

Abject, 59, 193
Adorno, Theodor, 18, 72
Aesthetic, 193
Aether, 211
Agapic love, 171
Agency: challenges of, 164–66; shared, 121–23
"Agential realism," 49, 80
Agnostic, 52, 63–84, 193
Agrippa's trilemma, 205, 212; circularity, 19, 21, 31, 194; foundationalism, 19–20, 21, 194, 201–2; identities and ethics in, 21; identity trilemma and, 86–102; infinite regress, 20; knowledge claims in, 18–19, 20
Alhambra, 197
Althaus-Reid, Marcella, 58, 209
Anarchism, 20, 194
Anatheism (Kearney), 67
Anekanta, 33, 83, 194
Anemic humanism, 159, 160

Animistic, 194
Anthropic principle, 194
Anthropocentrism, 35, 194, 204; epistemic, 200–1; knowledge-based, 200; ontological, 200–1
Anthropomorphism, 156
Antifoundationalism, 210
Apocalypse Now and Then (Keller), 131
Apocalyptic discourse, 195
Apocalypticism, 195
Apocalyptic thinking, 131
Apophatic theologies, 67–68, 70, 131, 194–95, 208
Apparatus, 40, 158, 159, 178n11
A priori, 195
Arboreal thinking, 46, 195, 212
Arche, 194
Archimedean standpoint, 195
Archimedes, 195
Archipelagic self, 195
Ardipithicus, 98, 99–100

228 INDEX

Aristotelian teleology, 27–28, 201
Aristotle, 19, 196, 213; universe of, 27, 211
Assemblages, 46, 51, 130, 153, 195–96
Athanasius, 197
Atheism, 9, 10, 11, 22, 78, 81, 193
Atomism, 203, 209
Autopoeisis, 52, 71, 196

Barad, Karen, 151, 174n13; on knowing, 57; on matter, 3–4, 40, 55; on nature, 38; on observer, 24; on performativity, 49, 157
Bauman, Zygmunt, 66, 114, 115, 117, 132, 146
Becoming: God/nature, 44; human, 14, 109, 151, 152; nature, 38; planetary, 8, 12, 34, 116–17, 154, 155, 156
Becoming plant, mineral, animal, 154–59
Being: knowing and, 201; reality and, 208
Being Human (Peterson), 12
Bennett, Jane, 123, 151
Bergson, Henri, 194, 199–200; *élan vital* and, 47–48; as emergentist, 47–48, 50
Beyond Good and Evil (Nietzsche), 17
Biohistorical creatures, 75, 76, 79, 109, 159
Biological evolution, 48, 75
Bisexual, *see* Lesbian, gay, bisexual, transgender, queer or questioning
Bodies That Matter (Butler), 57, 107
Bodies without organs (BwOs), 162–63
Boelstorff, Tom, 96, 98, 99, 195
Bohr, Niels, 38, 157, 215
Bottom-up causation, 71, 72
Boundaries: of species, 151, 154; of subjectivity and humanities, 155, 156
Boundedness, conceptual, 2
Brennan, Teresa, 113–14, 132, 136

"Bretton Woods" institutions, 136, 199
Bruno, Giordano, 8, 41, 42, 67, 207; multiverse of, 43–44, 45, 179nn20–21
Buddhism, 7, 123; concepts of, 9, 53, 54, 68, 69, 90, 104; dependent coarising in, 54, 68; emergence of, 82; tenets of, 9, 196, 210
Buddhists, 11, 34, 39, 60, 67, 68, 80
Butler, Judith, 57, 107, 108, 209, 215
BwOs, *see* Bodies without organs

Capitalism, 100, 185n35, 199, 209; free-market, 136; identity construction politics of, 102–5; secular myth and, 31–36; technocapitalism, 213
Capra, Fritjof, 55
Cartesian Cogito, 18, 19, 30, 94, 121
Cataphatic theology, 196
Categorical imperatives, 195
Catholic Church, 208
Causality, 122, 196
Centralization, 132–33, 136
Centrism, 203–4
Chinese culture, 197, 202
Christianity, 197, 208; concepts of, 28, 82, 83, 90; gay marriage and, 88; narrative of, 20; spread of, 60; theology of, 27, 28, 29
Christians, 11, 28, 29, 80, 88, 103, 197, 202
Circularity, 19, 21, 31, 86, 194
Cisgendered, 95, 99, 196
Classism, 122–23
Clayton, Philip, 71
"Clear and Distinct Ideas" (Descartes), 19
Climate change, 4, 10, 11, 14, 146
Coconstruction, 1, 11, 22, 49, 56, 107, 197
Colonization, 11, 36, 141, 210
Common grounds, 5, 135, 137–38, 174n8

Companion species, 121, 157
"Conceptual boundedness," 2
Conceptual conformity, 35, 177n42
Conceptual thinking, 3
Conceptual violence, 2, 15, 64
Confucianism, 7, 23, 39
Connolly, William, 32, 54, 73, 127
Conscious identities, 142
Consciousness, 90, 200
Constructions: ecosocial, 9; identity, 102–5, 193; of knowledge, 197; theological, 75, 182n30
Constructive-relative position, 197
Constructivism, 197
Contextualism, 73, 197, 212
Convivincia, 28, 197
Copernican Revolution, 197; *see also* Galilean Revolution
Copernicus, 42, 197, 202, 211
Cordoba Institute, 29
Cordova, 197
Cosmic expansion, 56, 200, 201
Cosmology, 26–31, 36, 54, 200
Cosmos, 28, 176n28, 201, 203
Creatio ex nihilo, 76, 197–98
Creation: continuous, 77; Keller on, 78, 110; stories of, 38
Creative-destructive life process, 25, 176n24
Creatures: biohistorical, 75, 76, 79, 109, 159; experiential, 51; meaning-making, 152, 153; response-able meaning-making, 107; response-able planetary, 118–19
Crossing and Dwelling (Tweed), 117, 138
Cultures: Chinese, 197, 202; human, 75; Indic, 197, 202; Vedic, 69, 82–83, 105; Western, 24, 213

Cyborg, 159–60
Cyborg ontology, 5

Dadaism, 20, 198
Daoism, 23
Dark Ages, 28, 38, 196, 202, 212
Deacon, Terrence, 50, 51, 52–53, 146
Death: of God, 63–84, 198; of nature, 7, 8, 40–41, 54
"The Death of Environmentalism" (Shellenberger and Nordhaus), 145, 190n46
The Death of Nature (Merchant), 8, 26
Deconstructionism, 20, 120, 198, 210, 213
Deleuze, Gilles, 52, 110, 134, 154, 195, 213; assemblages and, 51, 196; BwOs and, 162–63; as creative thinker, 10, 156, 198, 201; lines of flight of, 143, 206; rhizomes and, 46, 53, 78, 212
Dependent coarising, 54, 68–69, 196–97
Derrida, Jacques, 6, 20, 131, 198, 202
Descartes, 19, 30, 36, 195
Destabilizing identity: capitalist politics, of identity construction, 102–5; "I," 85, 86–102; identity solipsism, 86–102; identity trilemma, 86–102
Destabilizing nature, 37–61
Destabilizing religion, 63–84
Deterritorialize, 51, 135, 198–99, 213
Dialectic of Enlightenment (Horkheimer and Adorno), 72
Dialogical interaction, with religion and science, 25–26
Différance, 35, 59, 199
Dimorphism, 57, 88, 95–96, 97, 202, 205, 213
Divine love, 170
Divine power, 46
Divine promiscuity, 84

Dominion, 194
Dominology, 199
Dualism, 71, 200, 203, 205, 206

Earth others, 151
Eastern traditions, 24, 53
Ecological footprint, 115, 187n19
The Ecological Thought (Morton), 120, 161
Ecology nomadology, 137
Ecology Without Nature (Morton), 135
Economic globalization, 2, 108, 112, 132–33, 199
Economization, 2, 66
Ecoreligious identities, 107–25
Ecosocial constructions, 9
Ecosocial disparities, 114–15
EcoSpirit (Keller and Kearns), 136, 137
Ecosystems, 14, 49, 75, 199, 211
Ecotone, 199
Efficient causality, 196
Einsteinian physics, 48, 54
Élan vital, 47–48, 50, 194, 199–200, 206
Emergence: of ecoreligious identities, 107–25; evolutionary, 146; of human culture, 75; theory of, 42, 50, 51, 71, 122, 158, 200
Emergentists, 196, 200; Bergson as, 47–48, 50; Spinoza as, 8, 40, 41, 42
Emergent meaning-making practices, 71–74, 78, 182n23
Emergent newness, 51, 71, 72, 76, 153
Emergent phenomena, 9
Emptiness, form and, 68–69
End-oriented solutions, 213
Energy: dissolution of, 200; transfer of, 199, 200
Enfolding, 110, 111, 200

Enframing, 22, 110
Enlightenment, 18, 36
Ensoulment, 103
Entropy, 200
Environmental crisis, 8, 134, 135, 136
Environmental ethics, 13, 14, 108, 141, 194, 209, 211
Environmentalists, 59, 131, 134, 158–59
Environmental justice movement, 146
Epistemic anthropocentrism, 200–1
Epistemology, 20, 25, 55, 77; global, 112–13; of infinite regress, 73; planetary, 34, 116, 117; postfoundational, 116; of representation, 158; as study of knowledge, 18; theological, 79; uncertainty of, 132
Epistem-ontology, 55, 127, 130–35, 187n2, 201
Equilibrium, 208
Erotic love, 170
Essentialism, 201, 212
Essentialist thinking, 197
Eternity, 152
The Ethical Slut, 91
Ethics: in Agrippa's trilemma, 21; basis of, 12; environmental, 13, 14, 108, 141, 194, 209, 211; beyond exceptionalism, 12–15; global, 153; of movement, 209; place-based, 13, 127, 141, 209; truth, actions, and, 211; *see also* Planetary environmental ethics
The Ethics of Mobilities (Bergmann and Sager), 129, 188n6
The Ethics of Uncertainty (Anker), 1
Evangelical thinking, 202
Event-based metaphysics, 208
Event-based ontology, 201, 205, 212

Evolution: biological, 48, 75; knowledge and, 201; planetary, 60; process of, 52, 200
Evolutionary emergence, 146
Evolutionary rainbow, 89
Experiential creatures, 51

Feuerbach, Ludwig, 76–77, 81, 183n41
Filial love, 170
Filmer, Robert, 35–36
Final causality, 122, 196
Formal causality, 122, 196
Forum on Religion and Ecology, 8, 135, 143
Foucault, Michel, 3, 100, 202, 211, 215
Foundational-based identities, 12
Foundational fantasy, 132–33, 141
Foundational identity markers, 98
Foundationalism, 19–20, 21, 79, 86, 88, 194, 201–2
Foundational thinking, 1, 46, 134, 197, 214
Free higher education, 147–48
Free-market capitalism, 136
Fundamentalism, 202
Fundamentalist thinking, 202

Gaard, Greta, 87
Galilean Revolution, 202; *see also* Copernican Revolution
Galileo, 42, 197, 202, 211
Gay and Gaia (Spencer), 158
Gays: marriage of, 86–93; *see also* Lesbian, gay, bisexual, transgender, queer or questioning
Gedanken, 4, 55, 174n7
Gender dimorphism, 57, 88, 95–96, 97, 202, 213

Genders, 2, 196, 202; third/fourth, 3, 58, 88, 93–97, 213–14
Genesis, 35–36
Genuine, 102
Globalatinization, 31–36, 129, 188n10, 202
Global climate change, 4, 10, 11, 14, 146
Global epistemology, 112–13
Global ethic, 153
Globalization, 3–4, 11, 202, 209; economic, 2, 108, 112, 132–33, 199; of family values, 102–3; of free-market capitalism, 136; God of, 145; grounds for, 26–31; ill effects of, 10; logic of, 60; monological logic of, 18; planetary technologies and, 6; polarity and, 15; religious ideas with, 130, 188n12; Western, 199
Globalization (Bauman), 66, 114, 115
Globalization and Its Terrors (Brennan), 113
Global mobiles, immobile locals and, 114–15, 132
Global movement, of resources, 136–37
Global speed, 113–14
Glocal, 202
Glorious Revolution, 35
God, 76–77, 109, 146; death of, 63–84, 198; of gaps, 172; as given or natural, 198; of globalization, 145; image of, 10, 30, 81; knowing of, 196; omni-God, 67, 81, 133; reality of, 194–95; thinking about, 211; worldview of, 195
God/nature becoming, 44
Golden Age of Islam, 7, 28, 197, 198, 202, 212
Goodchild, Philip, 27, 65
Grant, Ian Hamilton, 42, 63
Green revolution, 203

232 INDEX

Grounds, 79, 183nn46–48; common, 5, 135, 137–38, 174n8; for globalization, 26–31; Keller on, 116, 136, 137; of life, 128
Guattari, Felix, 52, 110, 134, 154, 195, 213; assemblages and, 51, 196; BwOs and, 162–63; as creative thinker, 10, 156, 198, 201; lines of flight of, 143, 206; rhizomes and, 46, 53, 78, 212

Habits, of nature-culture, 48–54, 55, 108
Hadron Collider, 56, 215
Haraway, Donna, 5, 57, 71, 121, 156, 201
Hefner, Philip, 109
Hegelians, 107, 204
Heidegger, Martin, 22, 65, 109, 110, 175n12
Heise, Ursula, 112, 127, 140
Heliocentric universe, 197, 203
Heraclitus, 47, 201
Hermaphrodites, 58
Hernandez, Albert, 70
Heteronormativity, 90, 97, 100, 141, 203
Heterosexism, 88, 91, 140, 203, 209
Heterosexual identities, 88
Hinduism, 7, 11, 23, 38, 53, 82, 104
Historicity, 75
Holism, 203–4, 209
Homo sapiens, 100, 120, 156
Homo sapiens sapiens, 11, 41, 52
Homosexuality, 194
Horizontal meaning making, 138, 139, 140
Horkheimer, Max, 18, 72
How the Hippies Saved Physics (Kaiser), 55
Human becoming, 14, 109, 151, 152
Human culture, 75
Human exceptionalism, 204; challenges of, 151–72

Humanism, 154–55; anemic, 159, 160
Humanity, 13, 152, 155
Humans: as *homo sapiens sapiens*, 11, 41, 52; in hybrid relationships, 152; understanding of, 6, 12; at universe center, 203
Hybrid/hybridity, 1, 5, 11, 97, 152, 159–60, 173n1, 204
Hybrid narrated beginnings, 5
Hyperobject/hypersubject, 14, 120–21, 160, 161–64, 204
Hypocrisy, 164, 166–68

"I," 85, 86–102, 120
Iconoclasm, 204
Ideal and material realms, 87, 184n3
Idealism, 71, 75, 86, 200; Hegelian, 204; identity, 93–97, 101
Identities: in Agrippa's trilemma, 21; conscious, 142; ecoreligious, 107–25; formation of, 26, 39, 57, 59, 83, 84, 86, 101, 102, 105, 186n40, 197, 199, 205, 210; foundational-based, 12; heterosexual, 88; individualistic, 196; as material-ideal, 5, 55; multiple, 2; origins of, 1; performance of, 209; place-based, 131; planetary, 1–15, 123–25, 135; substance-based notions of, 1, 101; understanding of, 1; version of, 209; Western, 195
Identity-based political struggle, 90
Identity construction, 102–5, 193
Identity idealism, 93–97, 101
Identity markers, 98
Identity materialism, 97–102
Identity monism, 86–93, 101
Identity politics, 13, 93
Identity solipsism, 86–102

Identity trilemma, 86–102
Idolatry, 195
Immanence, 33, 204–5, 211; metaphysics of, 75, 182n29; radical, 38, 39, 40, 42, 48, 49, 152, 179n37
Indic cultures, 197, 202
Individualism, 209
Individualistic identities, 196
Individuality, 204
Indonesia, 96–97, 104
Indra's net, 54, 205
Infinite regress, 20, 73, 86, 134, 194, 205
Instrumentalization, 205
Instrumental thinking, 205
Integrity, 164, 168–69
Interdisciplinary knowledge production, 79
International Monetary Fund, 199
Intersexed, 13, 58, 93–97, 205, 206
Interstitial, 205
Into the Cool: Energy Flow, Thermodynamics and Life (Schneider and Sagan), 55
Irenaeus, 197
Islam, 104

Jainism, 4, 33, 90; concepts of, 7, 104, 194, 213; self in, 104; of Vedic background, 82–83
Jealousy, 91
Jesus of Nazareth, 68
Jews, 29
Judaism, 82

Kaiser, David, 55
Kant, Immanuel, 195
Kaufman, Gordon, 71, 75, 76, 80
Kearney, Richard, 67
Kearns, Laurel, 136, 137

Keller, Catherine, 209; on apophatic tradition, 67, 131; on creation, 78, 110; on grounds, 116, 136, 137
Knowing, 57; being and, 201; of God, 196; *see also* Planetary knowing
Knowledge, 195, 196; anthropocentrism based on, 200; construction of, 197; life evolution shaped by, 201; pragmatism and, 210; reality and, 211; regimes of, 3, 63
Kung, Hans, 153

Language, 160, 191n20
Latin West, 198, 212
Latour, Bruno, 135; on abject, 59; on religion, 63, 138; on transcendence, 33; on truth, 48, 143–44, 190n45
Laws of gravity, 13
Legal system, 122–23
Leisure, play, meditative space-time, 148–50, 190n48
Lesbian, gay, bisexual, transgender, queer or questioning (LGBTQ), 88–89, 96–97, 143, 189n29, 203, 205–6
Life, 132, 137, 138; cycle of, 129, 131; flourishing of, 135, 139; ground of, 128; process of, 134; vision of, 130
Lines of flight, 10, 82–84, 143, 206
"Liquid modernity," 117
Locke, John, 35–36
Lockean self, 85, 86, 93, 94, 121, 141, 163, 190n41
Logocentrism, 20, 206
Loves, 92, 164, 170–71, 209; multiple planetary, 169–72; of place, 131

Marx, Karl, 100
Marxist materialism, 39, 107, 204, 206

Material causality, 196
Material conditions, 132
Materialism, 71, 85, 86; identity, 97–102; Marxist, 39, 107, 204, 206; radical, 48–54, 155, 205; reductive, 33, 200, 202, 206
Matter, 55, 108, 157; design of, 52; importance of, 39; production of, 3–4, 40
McKibben, Bill, 41
Meaning, 186n10; in dark, 64–67, 181n3; of technologies, 109–12, 114, 213; understanding of, 138; *see also* Performing meaning: taking on abject toward planetary identities
Meaning-making contexts, 75–82
Meaning-making creatures, 152, 153
Meaning-making practices, 48, 107, 108, 111, 144; emergent, 71–74, 78, 182n23; horizontal, 138, 139, 140; polytheistic mode of, 139; religion related to, 11–12, 22, 40, 68, 84, 137, 138; science relating to, 39, 56, 60, 68
Mechanism, 206
Medieval Period, *see* Dark Ages
Meditative space-time, 148–50, 190n48
Meeting the Universe Halfway (Barad), 151
Merchant, Carolyn, 8, 26, 41, 57, 174n15
Metaphysics, 42, 206; event-based, 208; of immanence, 75, 182n29; nonsubstance, 208; process-based, 208, 211; substance-based, 7, 54, 196–97, 212
Mind over matter, 93–97
Mineral, 154–59
Misplaced concreteness, 37, 177n1
Mobility, movement and, 129
Modernity: place-less-ness of, 209; postmodernism and, 210, 213, 215

Modern science, 6–7, 30–31, 40, 98; revolution of, 197
Monism, 71, 86–93, 206
Monogamous pair bond, 100, 142
Monogamy, 90–91, 92, 141–42, 209
Monological logic, 206–7; of globalization, 18
Monotheism, 7, 31
Monotheistic nomadism, 139
Morton, Timothy, 14, 38, 120, 135, 161–62
Multinaturalism, 207
Multiperspectivalism, 33, 35, 82, 83, 111, 117, 207, 208, 213
Multiple histories, 1–2
Multiple identities, 2
Multiple loves, 209
Multiple natures, 34, 55
Multiple planetary loves, 169–72
Multiple subject, 119–20
Multiple technologies, 111–12
Multiscalar, 25, 207
Multiverse, 43–44, 45, 56, 179nn20–21, 207
Muslims, 11, 23, 29, 104, 111, 197, 202
Myth, of origins, 97–102

Nancy, Jean-Luc, 51, 209
Narrativity, 72
Natural-cultures, 37, 40, 42, 48–54, 71, 81, 137
Naturalistic fallacy, 101
Natural sciences, 37
Natura naturans, 25, 38, 44–47, 176n23, 179n23, 179n24, 179n30, 207–8
Natura naturata, 207–8
Nature: concept of, 12, 25, 207; death of, 7, 8, 40–41, 54; definition of, 25, 134–35; destabilizing, 37–61; ecology without,

140; multiple, 34, 55; place-based, 131, 135; queering, 56–61; of reality, 201, 211; of truth, 107; understanding of, 136
Nature becoming, 38
Nature-culture, habits of, 48–54, 55, 108
Nature naturing, 48, 135; see also Natura naturans
Negative theologies, 208
Neti-neti, 9, 208
New Atlantis (Bacon), 30
Newness, 51, 71, 72, 76, 153
Newton, Isaac, 30, 36, 206, 211
Newtonian mechanical universe, 212
Nicolas of Cusa, 43, 44, 45, 67
Nietzsche, Friedrich, 17, 67, 198, 202
Nomadic organism, 151
Nomadic polyamory of place, 127
Nomadic thinking, 79, 183n46
Nomadic understanding, polytheistic, 139, 141, 143
Nomadism: monotheistic, 139; planetary, 14, 136, 138, 189n27
Nomadology, ecology, 137
Nondualism, 53
Nonequilibrium: open evolving systems and, 54–56; state of, 55
Nordhaus, Ted, 145
"Nothing from nothing," 197

Objective-style thinking, 197
Observer, Barad on, 24
Occupy movement, 145
Omega Point, 38, 178n3
Omni-God, 67, 81, 133
One-fifth world, 4, 18, 65, 132, 173n6, 188n18

On the Infinite Universe and Worlds (Bruno), 43
Ontoepistemology, 158
Ontological, 208
Ontological anthropocentrism, 200–1
Ontological vegan, 155, 190n40
Ontology: of closet, 101; cyborg, 5; vepistem-ontology, 55, 127, 130–35, 187n2, 201; event-based, 201, 205, 212
Open evolving systems, 54–56
Opiate of masses, 139
Orientalism, 39, 43
Orientalism (Said), 54, 180n50
Origins: of identities, 1; of myths, 97–102
Orthodox Church, 208
Orthodoxy, 103, 208–9

Pantheism, 44–47
Particularity, 78
Paticcasamuppāda, 210–11
Patriarchy, 209
Performance, of identities, 209
Performativity, 40, 49, 58, 124, 157, 209
Performing meaning: taking on abject toward planetary identities, 123–25
Per/version, 58, 209
Peterson, Anna, 12
Pharmaceuticals, 6, 214
Phenomena, 55–56
Philosophies of Nature After Schelling (Grant), 42, 63
Philosophy and Social Hope (Rorty), 37
Physicists, 55
Physics, 48, 54; quantum, 157, 214–15
Place: illusion of, 133–34; love of, 131; nomadic polyamory of, 127; understanding of, 133, 137

236 INDEX

Place-based ethics, 13, 127, 141, 209
Place-based identities, 131
Place-based nature, 131, 135
Place-based politics, 143
Place-based technologies, 142
Place-based theories, 129
Place-based thinking, 127, 129, 210
Place-less-ness, 209
Planetarity, 108, 116, 186n1, 209
Planetary becomings, 8, 12, 34, 116–17, 154, 155, 156
Planetary community, 91
Planetary environmental ethics: anthropological/political level, 127, 140–44; components of, 144–47; epistem-ontological level, 127, 130–35, 187n2; ethical-religious level, 127, 135–40; free higher education, 147–48; leisure, play, meditative space-time, 148–50; planetary flows: context of movement, 128–30; universal health care, 148
Planetary epistemology, 34, 116, 117
Planetary evolution, 60
Planetary flows, context of movement, 128–30
Planetary identities, 1–15, 123–25, 135, 151
Planetary ills, 146
Planetary knowing, 18–22; and becoming, 116–17
Planetary nomadism, 14, 136, 138, 189n27
Planetary politics, 62
Planetary religions, 8–12, 18, 75–82
Planetary stories: identities and ethics beyond exceptionalism, 12–15; of religion, 8–12, 18; of science, 5–8
Planetary technologies, 3, 6, 15, 119
Planetary time, reverberations, 117–18

Plant, 154–59
Platonic forms, 9, 52, 195
Play, *see* Leisure, play, meditative space-time
Plumwood, Val, 93, 131, 141, 201
Plurisingular, 209
Polarity, 15
Political struggle, identity-based, 90
Politics: capitalist, 102–5; identity, 13, 93; place-based, 143; planetary, 62
The Politics of Nature (Latour), 135
Polyamory, 92, 127, 169, 209–10
Polydoxy, 14, 63, 64, 82, 83, 175n22, 210
Polytheistic mode, of meaning making, 139
Polytheistic nomadic understanding, 139, 141, 143
Positive theology, 196
Postcolonialism, 210
Post-Einsteinian physics, 48, 54
Postmodern constructivism, 94
Postmodernism, 118, 210, 213, 215
Postmodern sciences, 41, 178n16
Poststructuralism, 210
Pragmatists, 48–54, 210
Pratītyasamutpāda, 210–11
Precautionary principle, 134, 211
Prime Mover, 28, 176n27
Process-based metaphysics, 208, 211
Process thought, 211
Projections, of theologies, 68, 72, 75–81
Pseudo-Dionysius the Areopagite, 67
Ptolemaic cosmology, 26–31
Ptolemaic geocentric universe, 203, 211
Ptolemy, 211

Quantum level, 49, 215
Quantum physics, 157, 214–15

Quantum reality, 45–46, 54, 214–15
Quantum theory, 55
Queer, 86, 193; see also Lesbian, gay, bisexual, transgender, queer or questioning
Queering nature, 56–61
Queer theory, 89, 202, 209, 215; definition of, 21; nature and, 8, 13, 41; in planetary community, 158–59
The Question Concerning Technology (Heidegger), 109
The Quest for the Historical Satan (De La Torre and Hernandez), 70

Racism, 57, 99, 122–23, 165
Radical immanence, 38, 39, 40, 42, 48, 49, 152, 179n37
Radical materialism, 48–54, 155, 205
Radical orthodoxy, 210
Rationality, 80, 183n50
Reduction/reductionism, 72, 75
Reductive materialism, 33, 200, 202, 206
Reductive thinking, 1
The Re-Emergence of Emergence (Clayton), 71
Regimes: of knowledge, 3, 63; of truth, 61, 153, 164, 211
Reification, 37, 65, 211
Relative perspectives, 213
Relativism, 197, 211–12
Relativity, 116
Religions, 174n17; definition of, 22–23, 138; destabilizing, 63–84; dialogical interaction with science and, 25–26; with globalization, 130, 188n12; interpretations of, 12; as lines of flight, 10, 82–84, 206; meaning-making practices in, 11–12, 22, 40, 68, 84, 137, 138; planetary, 8–12, 18, 75–82; science and, 17, 25–26, 44, 60, 85; understanding of, 63–64; see also Ecoreligious identities; Ethical-religious level
Religions of the World (Tucker), 135
Renaissance, 7, 38, 198, 212
Representation, of epistemology, 158
Representational model, of reality, 158
Response-able meaning-making creature, 107
Response-able planetary creatures, 118–19
Responsibility, 74
Reterritorialize, 51, 135, 198–99, 213
Revelation, 109, 195
Reverberations, 74, 117–18
Revolutions: Copernican, 197; Galilean, 202; Glorious, 35; Green, 203; modern scientific, 197; scientific, 26–31
Rhizomatic thinking, 74, 78, 195
Rhizomes, 46, 53, 78, 212
Rorty, Richard, 37, 79
Rubenstein, Mary Jane, 20, 85, 86, 90
Ruether, Rosemary Radford, 21, 100

Sagan, Dorian, 55, 60
Said, Edward, 53–54
Salvation, 204, 205, 213
Satan, 70
Schizophrenia, 87, 88, 95
Sciences: definition of, 24–25; influence on, 39; meaning-making practices in, 39, 56, 60, 68; natural, 37; of planetary stories, 5–8; postmodern, 41, 178n16; religions and, 25–26, 44, 60, 85; role of, 37–38; see also Modern science; Western science
Scientia, 24
Scientific revolution, 26–31
Scripture, 28, 83, 88, 90, 208
Second Treatise (Locke), 35

A Secular Age (Taylor), 32
Secular myth, capitalism and, 31–36
Secular theologies, 205
Self, 197; archipelagic, 195; birth of, 81, 184n56; definition of, 74; in Jainism, 104; Lockean, 85, 86, 93, 94, 121, 141, 163, 190n41
Self-and-other, understanding of, 1
Self-organization, 52
Semeyana, Caster, 13, 97–102
Sense of Place, Sense of Planet (Heise), 112
Sex/gender, 2, 196, 202
Sexism, 57, 95, 122, 203
Sexual dimorphism, 57, 95, 205
Shared agency, 121–23
Shellenberger, Michael, 145
Shinto, 194
Shiva, 69–70
Simians, Cyborgs, and Women (Haraway), 5
Slippery slopes, 130–35
Solipsism, 19, 86–102, 212
Space: movements in, 152; of transcendence, 78
Space-time, 54, 73, 78, 129, 138, 144–48; meditative, 148–50, 190n48; political, 38
Spatial centralization, 136
Species: boundary of, 151, 154; companion, 121, 157; as nomadic organism, 151
Speculative realism, 205
Spencer, Daniel, 158
Spinoza, Baruch: as emergentists, 8, 40, 41, 42; *natura naturans* of, 25, 38, 44–47, 179n23, 179n24, 179n30, 207–8; pantheism of, 44–47
Spivak, Gayatri, 6, 112, 138, 209, 212
Stewardship, 194

Stories: of creation, 38; *see also* Planetary stories
Strange Wonder (Rubenstein), 85, 86
Strategic essentialisms, 212
Subjection, 58
Subjectivities, 98, 163, 168, 170–72; apparatus and, 40; assemblages and, 51, 152; boundaries and, 155, 161; definition of, 99, 120; performance and, 111, 125; shaping of, 97, 100, 166; understanding of, 99, 116, 119, 157, 162, 164, 165
Subjects: multiple, 119–20; positions of, 99
Substance-based identities, 1, 101
Substance-based metaphysics, 7, 54, 196–97, 212
Sui generis, 212
Sun, at Universe center, 203
Syadvada, 33, 83, 207, 213
The Symbolic Species (Deacon), 50

Taoism, 39, 52
Tao of Physics (Capra), 55
Taylor, Charles, 32, 33
Techne, 6, 109, 159
Technocapitalism, 213
Technologies, 151; of meaning, 109–12, 114, 213; multiple, 111–12; place-based, 142; planetary, 3, 6, 15, 119
Technology and Human Becoming (Hefner), 109
Tehom, 110, 186n8
TEK, *see* Traditional ecological knowledge
Teleology, 52, 213; Aristotelian, 27–28, 201
Territorialization, 213
Tertullian, 197

Theism, 10, 33, 193
Theological constructions, 75, 182n30
Theological epistemology, 79
Theological reasoning, 80, 184n51
Theologies: apophatic, 67–68, 70, 194–95, 208; cataphatic, 196; of Christianity, 27, 28, 29; negative, 208; positive, 196; projections of, 68, 72, 75–81; secular, 205
Theories, 60, 193; emergence, 42, 50, 51, 71, 122, 158, 200; of everything, 200; place-based, 129; quantum, 55; queer, 8, 13, 21, 41, 89, 202, 209, 215; of Western culture, 24, 213
There Is No Alternative (TINA), 199
Thermodynamics, 53, 200, 208
Third/fourth genders, 3, 58, 88, 93–97, 213–14
A Thousand Plateaus (Deleuze and Guattari), 154
Thus Spake Zarathustra (Nietzsche), 198
TINA, 199
"Top-down causation," 71, 181n17
"Toward a Queer Ecofeminism" (Gaard), 87
Traditional ecological knowledge (TEK), 7, 214
Transcendence, 33, 214; space of, 78; world and, 17
Transcendent reality, 11
Trans community, 95
Transcultural ecoreligious identities, 108, 119
Transgender, 95, 196; see also Lesbian, gay, bisexual, transgender, queer or questioning
Tricksters, 120, 214; as creative destroyers, 69–70
Trinity, 28

Truth, 98, 185n29; ethics and actions of, 211; Latour on, 48, 143–44, 190n45; nature of, 107; regimes of, 61, 153, 164, 211; as transcendent reality, 11; understanding of, 31
Tucker, Mary Evelyn, 8, 135
Turing Test, 157, 191n12
Tweed, Thomas, 23, 117, 138, 140

Unearned privileges, 122
United Nations Environmental Program, 143
Universal Declaration of Human Rights, 26, 153
Universal health care, 148
Universe, 13; of Aristotle, 27, 211; heliocentric, 197, 203; humans at center of, 203; Newtonian mechanical, 212; Ptolemaic geocentric, 203, 211; sun at center of, 203
Universe Story, 38, 178n4
Unmoved Mover, 27, 28

Vedic culture, 69, 82–83, 105
Vibrant Matter (Bennett), 123, 151
Violence, conceptual, 2, 15, 64

Waria, 96–97
Western civilization, 1, 8
Western cultures, 24, 76, 213
Western environment, 53, 131
Western globalization, 199
Western histories, 39, 109
Western identities, 195
Western science: religion and, 17; worldview of, 205, 206, 214
Western thinking, 2–3
Whitehead, Alfred North, 201, 211

Wholism, 203

World: accountings of, 17; immanent understanding of, 204–5; of knowledge, 205; multiple perspectives of, 213; one-fifth, 4, 18, 65, 173n6; of reality, 214; transcendence and, 17

A World of Becoming (Connolly), 54, 127

World Bank, 199

Worldly Wonder (Tucker), 8

Worldview, 211; of God, 195; of Western modern science, 205, 206, 214

Žižek, Slavoj, 37, 38, 40, 140, 152

GPSR Authorized Representative: Easy Access System Europe, Mustamäe tee
50, 10621 Tallinn, Estonia, gpsr.requests@easproject.com

www.ingramcontent.com/pod-product-compliance
Lightning Source LLC
Chambersburg PA
CBHW021941290426
44108CB00012B/923